INTRODUCTION

The Breakthrough

All of us, whether or not we realize it, are living through the early stages of one of the most important scientific breakthroughs in history.

From the birth of modern chemistry shortly before 1800 until just a few years ago, biologists have puzzled over the nature of life, seeming to make progress only at the far borders of the subject. Some were ready, in discouragement, to mark down the question of life and its mechanisms as insoluble, as something the human mind could never penetrate and understand.

And then came the remarkable decade of the 1940's. With the world convulsed by war, a kind of creative frenzy seized the scientists of the world. (This connection between war and human creativity has been noted before, but it has rarely been advanced as a good excuse for war.)

Biochemists had already learned how to use radioactive atoms in their researches on living organisms. They incorporated these atoms into compounds, which they were then able to follow through the body. But then, in the 1940's, such atoms became freely available, thanks to the nuclear reactor, and through their use biochemists have skillfully unraveled some of the complicated strands of body chemistry.

Biochemists also learned, in that decade, to separate complex mixtures by using just a sheet of absorbent paper, some common solvents, and a closed box. On the other hand, they also bent fearfully complex instruments to their purposes: electron microscopes that enlarged objects hundreds of times more than ordinary microscopes could; mass spectrographs that sorted out atoms one by one; and so on.

In that same decade, they took the first steps toward the actual delineation of the fine structure of the giant molecules that make up living tissue.

But *the* breakthrough came in 1944, when a scientist named O. T. Avery, with two of his colleagues, studied a substance that was capable of changing one strain of bacteria into another. This was deoxyribonucleic acid, commonly known as DNA.

To the layman, this discovery may not sound important. Nevertheless, it completely reversed several concepts that biologists and chemists had been taking for granted for a century. It launched in a new direction the investigation into the nature of life and brought about new research methods. The branch of science now called "molecular biology" came into its own.

In a matter of less than twenty years since, problems that once seemed insoluble have been solved, views that once seemed fantasy have been shown to be fact. Scientists have been competing in a race for achievement, and most of them have emerged winners.

The consequences are almost beyond calculation, for the cold, clear vision of modern science has been able to reach a deeper level of understanding of man than at any time in its three and a half centuries.

Science as we now know it began about 1600, when the great Italian investigator Galileo popularized the procedure of applying quantitative methods to observation, of making accurate measurements, and of abstracting generalizations that could be expressed as simple mathematical relationships.

Galileo's victories were in mechanics, in the study of motion and forces. This field of inquiry was greatly advanced by an English scientist, Isaac Newton, toward the end of the seventeenth century. The motions of the heavenly bodies were interpreted according to the laws of mechanics; complex phenomena were deduced from basic and simple assumptions; and astronomy as well as physics began to take on a modern shape.

Physics continued to advance and flower along the lines of Galileo's original breakthrough. In the nineteenth century, electricity and magnetism were tamed, and theories satisfactorily accounting for electromagnetic phenomena were established.

With the opening of the twentieth century, the discovery of radioactivity, and the development of quantum theory and the theory of relativity brought physics to new heights of complexity and subtlety.

Meanwhile, at the end of the eighteenth century, the French chemist Lavoisier applied the methods of quantitative measurement to the realm of chemistry, and that field of knowledge became a true science. The nineteenth century saw the development of new and fruitful theories involving atoms and ions. Great generalizations were made: the laws of electrolysis were laid down, and the periodic table was established. Chemists learned to synthesize substances not found in nature, and these synthetics were sometimes more useful in specific ways than anything natural.

Toward the end of the nineteenth century, the division between chemistry and physics began to disappear. New branches of knowledge, such as physical chemistry and chemical thermodynamics, flourished. In the twentieth century, the quantum theory made it possible to determine the manner in which atoms joined to form molecules. Nowadays, any division between chemistry and physics is purely artificial: the two together make up a single science.

And while these great victories of the human mind over the inanimate universe were being won, while the physical sciences were growing into giants, what of the life sciences?

They did not stand still, of course; great advances were made. The nineteenth century, for instance, saw no less than three major breakthroughs.

In the 1830's the German biologists Schleiden and Schwann advanced the cell theory. In their view, all living things consisted of tiny cells, visible only under the microscope. These were the actual units of life.

In the 1850's, the English naturalist Darwin worked out a theory of evolution which bound all life, past and present, into a whole. That theory lies at the basis of modern biology.

Finally, in the 1860's, the French chemist Pasteur propounded the germ theory of disease. It was only then that physicians really began to know what they were doing, and medicine became more than a hit-or-miss-and-trust-in-God affair. It is from then that we date the drastic decline in death rates and the spectacular lengthening of life expectancy.

These breakthroughs in the life sciences, however exciting they may be, are not of the same nature as those in physics and chemistry: they are descriptive, qualitative; they do not involve the application of precise measurements. They are not generalizations of the type that make for confident prediction or for knowledgable manipulation of some facet of the universe.

This disparity in progress in the various fields of science has been the despair of many serious students of human affairs. As man's understanding of the universe about him has deepened and strengthened, the power in his hands has increased constantly. From gunpowder, he has gone on to high explosives and nuclear bombs. He has discovered new poisons, both chemical and biological. He even has a new "death ray," in the form of an instrument called a laser, which, however, also promises great things in the field of communications, industrial technology, even medicine, if we can but learn to concentrate on its peacetime uses.

Man has always been capable of using his knowledge to bring about pain and misery—he has had this power ever since he learned to use fire and picked up his first club. With the decade of the 1940's, however, he became for the first time capable of using his knowledge in such a way as to destroy the human race and perhaps all of life.

Science has been able to place this great knowledge at the disposal of human beings, but the human being himself remains beyond the understanding of science.

Now, what of the "social sciences"? Great minds have carefully studied the psychological drives, both "normal" and pathological. Others have studied the societies and cultures built by man. However, neither psychology nor sociology has yet done more than touch the very fringes of the subject or advance far beyond the stage of mere description. None of the social sciences is what a chemist, physicist, or physiologist steeped in quantitative measurements would call "science"; with the best will and effort in the world, the social scientists have still not discovered "what makes Sammy run."

So we come face to face with this fact: man knows enough now to kill a billion men in a single day by an act of his will; but he is not yet capable of understanding what lies behind that act of will.

"Know thyself," warned Socrates, 2,500 years ago. And now mankind had better learn to know itself, or we are all doomed.

To be sure, the physical sciences have encroached upon the territory of biology, lopping off a border area here, penetrating a short way inward there. Physicists have studied muscle contraction and the electrical potentials of the brain. Chemists have tried to work out the chemical reactions going on within living tissue. Most of the realm of biology, however, remained far out of reach, and the physical scientists were forced to chip away at the periphery only—until the great decade of the 1940's.

Then, in 1944, almost at a stroke, the central problem of life—of growth, reproduction, heredity, the differentiation of the original egg cell, perhaps of the very workings of the mind—was exposed to the scalpel of the physical sciences.

It was only then that the foot of man was for the first time placed upon the highway of true life *science,* a highway that may (and should) end finally in as detailed an understanding of life and mind as now exists in connection with atoms and molecules.

This new understanding, of course, might be abused, might serve as the source of a new horror: the scientific control of life could be bent to serve the purposes of a new tyranny. Then again, it might not: properly used, it could end or, at the very least, control most of the ills flesh is heir to, both physical and mental. Further, it could place the deadly forces of nature in the hands of a species that understands itself and can control itself—and can therefore be trusted with questions of life and death.

Perhaps it is too late; perhaps the insanity of man will drive us all to destruction before the new knowledge can reach the necessary level of excellence. But at least we can now make a race for it.

And perhaps we need only hold out for a generation or two, for the speed of advance in modern science is astonishing.

Consider—

In 1820, a Danish physicist named Oersted noticed that a compass needle swung when brought near a wire through which an electric current was flowing. This accidental observation first knit together the phenomena of electricity and magnetism.

It was a simple observation, and scarcely anyone could have foretold the consequences. As a result of the research stemming from what Oersted had noted, electrical generators and motors were developed and the telegraph was invented, all within a quarter of a century. Within 60 years, the incandescent light was invented, and the electrification of the world was begun.

In 1883, Thomas Edison observed that if a metal plate were sealed into a light bulb near the heated filament, an electric current could be made to travel across the vacuum between the filament and the plate in one direction but not in the other.

Edison himself did not recognize the value of this breakthrough, but others eventually did. The "Edison effect" was put to use in what are now called "radio tubes," and the science of electronics came into being. Within 40 years, radio had become a new force in human affairs. Within 60 years, television was replacing radio, and electronics was being applied to the building of gigantic computers.

In 1896, the French physicist Becquerel noted that a photographic film was fogged by the presence nearby of a uranium compound, even when the film was wrapped in black paper. Uranium was giving off penetrating (but invisible) rays, it seemed, and that observation opened up to science a new world within the atom.

Within a quarter of a century after Becquerel's discovery, atomic scientists were smashing atoms; within another quarter of a century, they were smashing cities. Within 60 years, nuclear stations were supplying power for civilian needs, and physicists were hot on the trail of man-made fusion power that may solve our energy needs for millions of years to come.

In 1903, the Wright brothers flew the first heavier-than-air flying machine. It was little more than a large kite with an outboard motor, and it managed to hop a few feet through the air, coming down again after a few seconds. In 60 years, however, that first airplane has given rise to descendants in the form of tremendous jetliners capable of carrying over 100 passengers across oceans and continents at more than the speed of sound.

In 1926, Goddard shot a rocket into the air, the first rocket ever to be powered by liquid fuel and liquid oxygen. It reached a height of 184 feet and a speed of 60 miles an hour.

But the technique of rocketry advanced rapidly and, within 35 years, rockets were developed that were capable of placing men into orbit about the earth, at distances of more than 100 miles, and at the speed of 18,000 miles an hour. In another quarter century, it seems beyond doubt, man will reach the moon and establish scientific bases there.

Sixty years, then, seems to be the typical time interval from scientific breakthrough to full flower. Since scientists studied in 1944 a substance which they called DNA, and since this undoubtedly revolutionized the life sciences with full breakthrough force, I feel confident that—if we survive—the year 2004 will see molecular biology introducing triumphs that can now barely be imagined. Many of us will survive to see that. And if we reach 2004 in safety, man may then be knowledgeable enough to guarantee his own safety even against the possibility of self-destruction.

This book is an attempt, then, to explain the background of the breakthrough; the full meaning of the breakthrough and its immediate consequences; and, finally, a forecast of what the breakthrough may bring about in the future—what the world of 2004 may be like, as seen through my own wishful eyes.

CHAPTER 1

Inheritance and Chromosomes

Before Science

Every woman knows when she becomes a mother. She knows that the baby is her own because it has issued forth from her own body.

The concept of fatherhood is a little harder to grasp. It took time before primitive man realized that he had any essential part in the creation of a baby. Eventually, however, this notion was grasped, and by the time any group of men had become civilized the idea of fatherhood was established.

Once the realization of fatherhood came, the family itself took on a new meaning. A child was no longer something that happened inexplicably to a woman, proving an encumbrance to the man who happened to be consorting with her at the time. It had also become part of the man himself: a portion of his own body come to life and grown young again.

From an encumbrance, the child became a symbol of immortality: a creature who would live on when the father died and who would remain the representative of the family. The child was a portion of a continuing group, whose feats redounded to the credit of all other members of the group, living and dead and yet to be born, and whose shame disgraced all. (The Bible itself has many references to the tragedy of childlessness, which would involve the death of a family.)

Along with the realization of fatherhood, there arose, almost inevitably, some feeling about the inheritance of qualities and characteristics. In the first place, the son often visibly resembled his father. This became, in fact, the one sure sign that the husband of the mother was indeed the father of the child.

From this it is but a step to the feeling that the son inherits the more intangible qualities of the father: courage, temperament, various skills. If a man has proven himself fit to rule, it can easily be assumed that his son must automatically possess those qualities that made the father fit to rule. It made sense, therefore, that kingship should descend from father to son.

It was this notion of a tight, organic unity, binding the generations through physical continuity and the inheritance of characteristics, that formed the background to the emergence of such phenomena as ancestor worship, feuds and vendettas, aristocracies, caste systems, and even racism.

Nor is this notion of family lost among us today. Many of the strict tribalist notions of early man have gone, but we still know exactly what we mean when we say that someone comes "of good family." We are still ready to visit the sins of the parents upon the children by suspecting that children of parents who "came to no good" can scarcely come to good themselves.

The notion, then, of the inheritance of characteristics, of the passing on of qualities from parents to children, is one of the oldest, most widespread, most firmly fixed notions held by the human race. It is one of the most important, too, considering the manner in which it has affected the human social structure.

Anything that can clarify the manner in which such inheritance of characteristics comes about, anything which reduces it from intuitive tradition to precise scientific knowledge, cannot fail to be of the greatest interest and importance.

Genetics

It was not until the 1860's that anyone actually experimented with the mechanism of inheritance. It was only then that precise observations were made, recorded, and carefully analyzed. The man who did this was an Augustinian monk named Gregor Mendel who, in an Austrian monastery, indulged in his hobby of botany. He grew pea plants of different varieties, crossed them carefully, and noted the manner in which different characteristics of coloring, of seed appearance, of length of stem were developed. Out of this experimentation came certain simple conclusions that are now called "the Mendelian laws of inheritance." These laws, it turned out, applied with equal vigor not only to peas, but to all creatures, to fruit flies, to mice, and to men.

When these laws were applied to human inheritance, it was deduced that both parents, male and female, contributed equally to the inheritance of the child. For each physical characteristic, each parent contributed (under the simplest circumstances) one factor. The two factors that thus governed a particular characteristic might not be quite alike. For instance, one parent might contribute an eye-color factor that produced blue eyes, while the other might contribute a brown-eyes factor.

In combination, one factor might outweigh the other. For example, a person who had inherited a factor for brown eyes and one for blue eyes would be brown-eyed. However, the factor for blue eyes would persist and, in combination with another such factor, might produce a blue-eyed child in the next generation.

In the early twentieth century, these factors came to be called *genes,* from a Greek word meaning "to give birth to," and the science dealing with the manner in which these genes are inherited, and in which the characteristics that they determine are displayed, was named *genetics.*

In working with pea plants, Mendel was fortunate to have dealt with a simple organism, the breeding of which he could control. The various characteristics he studied were each determined by a single gene pair, and so he obtained useful results. In the more complex organisms, characteristics are likely to be the product of a number of genes working in cooperation. Furthermore, such genes may produce characteristics that are themselves affected by environmental conditions. It then becomes difficult to disentangle the threads of inheritance.

Among human beings, particularly, there are problems. Some characteristics, such as blood types, can be followed fairly well. Many others, even some as apparently simple as skin color, have complicated patterns of inheritance that have not been worked out. To be sure, "folk wisdom" often supplies answers that seem plausible, and on which racist theories are built for which men may be willing to die. Matters are neither as simple nor as bloody for the scientist, however.

To work out the complexities of inheritance, it is not enough to deal with the intact organism and to study only those characteristics that appear to the unaided eye and mind. This is something like trying to work out the rules of football by studying only a series of final scores. One can tell from the number of times those scores are multiples of six and seven that there must be some play that counts six points and another play that adds a seventh point. If one could listen to the shouting from the stands, one might be able to deduce the fact that there is a minimum playing time of one hour divided into two equal halves. To get farther along, however, one would have to watch an actual game.

Cell Division

In the latter half of the nineteenth century, biologists got down to the "actual game." They did this by taking up the careful study of the microscopic cells which make up all living matter.

Each cell is a drop of fluid (very complex in structure and chemistry), surrounded by a thin membrane and possessing a small central body called a *nucleus*.

The cell is the unit of life, and although an organism

may be made up of many trillions of cells, all the properties and characteristics of the organism can be traced back to the functions and activities of one group of cells or another, or some combination of groups. The color of a man's skin depends upon the activity of certain cells in the skin which manufacture a brownish-black pigment. The greater the ability of those cells to manufacture this pigment, the darker the man's skin. If a man suffers from diabetes, it is because certain cells in his pancreas fail to manufacture, for one reason or another, a particular substance.

We can reason like this indefinitely, and as we do so we cannot avoid the thought that, if we understood how the properties and characteristics of cells are passed on, we would thereby understand how the properties and characteristics of whole organisms are passed on. Thus, skin cells periodically divide so that two new skin cells exist where only one had before. Each of the new cells has precisely the capacity to produce the pigment that the parent cell had originally. How was that capacity preserved?

About 1880, a German biologist, Walther Flemming, studied the process of cell division carefully. He found that the nucleus contained material which could pick up a red dye that would make it stand out clearly against a colorless background. This material was therefore named *chromatin* (kroh′muh-tin), from a Greek word for "color."

During the process of cell division, the chromatin collects into pairs of threadlike bodies called *chromosomes* (kroh′moh-sohmz). Since these threadlike chromosomes play the essential role in cell division, the process was named *mitosis* (my-toh′sis), from a Greek word for "thread." At the crucial moment just before the cell actually divides, pairs of chromosomes pull apart. One of each pair goes to one side of the dividing cell, while the other goes to the opposite side. When cell division is completed, each new cell has an equal number of chromosomes.

Put this way, it might seem that each new cell had only one half the original number of chromosomes. This, however, is not so. Before the separation of chromosomes, each chromosome somehow forms a replica of itself. (This process is, therefore, called *replication*.) It is only after this doubling that the cells divide. Each new cell, consequently, has a full set of chromosome pairs, identical with the original set possessed by the parent cell. Each new cell is ready for a new division, at which time the process of doubling, followed by halving, will be repeated.

Since chromosomes are so carefully conserved in cell division and so carefully parceled out among the new cells, it seems natural to assume that it is these chromosomes that somehow govern the characteristics and functions of the cells. If the daughter cells possess all the abilities of the parent cell, it is because they possess either the original chromosomes of the parent or exact replicas of those chromosomes.

But can one be sure that, because the chromosomes carry within their structure the ability to determine the characteristics of a particular cell, they can also be responsible for the characteristics of a whole organism? The best argument in affirmative answer to that question is to point out that all organisms, however large and complex they may be when full grown, begin life as but a single cell.

This is true of the human being, for instance, who begins life as a fertilized ovum, produced by the union of an egg cell of the mother and a sperm cell of the father. The original egg cell is the largest cell actually produced by the human body of either sex. Even so, it is only 1/200 of an inch across, just barely visible to the naked eye.

Somewhere in that small object are present all the factors that represent the mother's contribution to the inherited characteristics of the child. Most of the material in the egg, however, is food, itself inert and nonliving. It is the nucleus of the egg, making up a very tiny portion of the whole, that is the truly living part; it is that which carries the genetic factors.

This may sound like no more than a guess, until we consider the father's contribution. The sperm cell contains no food to speak of; once it combines with the egg cell, it is the food supply of the latter that will have to do for the fertilized egg. Consequently, the sperm cell is much smaller than the egg cell. It is, in fact, only 1/80,000 the size of the egg cell. It is the smallest cell produced by the human body of either sex.

That tiny sperm cell nevertheless contains the entire contribution of the father to the child's inheritance. That contribution exactly equals the mother's.

The interior of the sperm cell consists almost entirely of well-packed chromosomes, one of each pair existing in human cells. There are twenty-three altogether. The egg cell, in its own nucleus, also contains twenty-three chromosomes, one of each pair present in the mother's cells.

In the formation of the egg cells and the sperm cells, we have the only case of the parceling out of chromosomes without prior replication. Egg cells and sperm cells therefore have "half-sets" of chromosomes. This situation is corrected when sperm cell and egg cell fuse together to form the fertilized ovum, which contains twenty-three pairs of chromosomes, one of each pair from the mother and one of each pair from the father.

It is common knowledge that mother and father contribute equally to the characteristics inherited by the child. While the mother's egg cell contains much besides chromosomes, and the father's sperm cell contributes nothing more than one half-set of chromosomes, it seems therefore an inescapable conclusion that the chromosomes contain the genetic factor, not only for individual cells, but for entire organisms, no matter how complicated.

To be sure, since one cannot suppose that there are only 23 different characteristics in the human body, no one has suggested that each chromosome determines only one characteristic. It is assumed instead that each chromosome is made up of a whole string of genes, each of which determines a different characteristic. One modern estimate is that each human chromosome contains somewhat more than 3,000 genes.

Beginning about 1900, it came to be realized, thanks to the pioneer work of a Dutch botanist, Hugo de Vries, that inheritance doesn't always proceed smoothly. Sometimes new characteristics arise that do not resemble those of either parent. This is now called *mutation*, from a Latin word meaning "change."

Mutations can be interpreted in the light of chromosome theory. Sometimes chromosomes are imperfectly shared in the course of cell division, and an egg or a sperm cell can end up with one chromosome too few or too many. The resultant imbalance would then be shared by all the cells in the body.

The serious consequences of such imbalances have been fully recognized only in very recent years, at least as far as man is concerned. The chromosomes are present within our cells in what appears to be such a complex jumble that it was only in 1956 that the correct count of 46 chromosomes to the human cell was established. (Before that, it had been thought that there were 48 to the cell.) New techniques were developed for isolating and studying the chromosomes and, by 1959, it was discovered that children who were born with a form of mental retardation called "mongolism" possessed not 46 but 47 chromosomes to the cell. Other more or less serious disorders are also being traced to abnormal numbers of chromosomes and to distortions of chromosomes arising during cell division.

Nevertheless, not all mutations can be traced to gross changes in the chromosomes. Many, indeed most, show up where there is no visible change in the chromosomes at all.

It would seem reasonable to assume that, in these last cases, changes in the chromosome have taken place, but at a level invisible to the eye, even when that organ is aided by the microscope. The changes must have taken place in the submicroscopic structure of the substance making up the chromosomes.

If this is so, then it becomes time to probe still deeper— that is, to enter the domain of the chemist. But before we can ask what chemical changes do take place in chromosomes, we must first ask: Of what chemical substances are chromosomes composed?

CHAPTER 2

Of First Importance

The Substance of the Chromosome

The chemical makeup of living tissue has been a problem that has concerned chemists for a century and a half, although the broad outline was laid down by the middle of the nineteenth century.

The major component of all living tissue is, of course, water—the same water that exists everywhere in the world about us. The remaining material, however, consists in the main of compounds that are distinctly unlike the substances so common in the inanimate world.

The substances of soil, sea, and air are stable, resistant to heat, and, for the most part, not flammable. The substances isolated from living tissue, however, are easily destroyed by heating. All are more or less flammable and, even if heated in the absence of air, so that they cannot burn, are nevertheless decomposed. They then give off vapors and change permanently in one way or another.

As a result, the substances isolated from living tissue (or from once-living tissue) were, as long ago as 1807, given a classification of their own: they were *organic substances,* because they were isolated from organisms. Material obtained from the inanimate world was naturally classified as *inorganic substances.*

By 1820, it had become customary to think of the organic substances as falling into one or another of three broad groups: the *carbohydrates,* the *lipids* and the *proteins.* In terms of substances most familiar to us, sugar and starch are carbohydrates, olive oil and butter are lipids, while gelatin and dried egg white are proteins.

By the mid-nineteenth century, it seemed quite clear that, of these three, proteins were the most complicated in structure and the most important in function. In fact, the very name "protein" is derived from a Greek word meaning "of first importance."

The complexity of protein structure is reflected in the fragility of the substance. (It doesn't always follow, but one would expect a tall, complicated house of cards to be easier to upset than a very small one.)

Carbohydrates and lipids can withstand treatment that proteins cannot, at least not without losing their ability to function as proteins. Most proteins in solution, for instance, when exposed to gentle heat, change permanently: the protein becomes insoluble, and can no longer perform whatever natural function is ordinarily its own. It is *denatured.*

A touch of acid can denature a protein; a touch of alkaline solution can do so. So can strong salt solutions and radiation. In the absence of all these factors, even shaking a protein solution into a froth will often suffice to denature it.

In fact, proteins seem the very stuff of life, as fragile and tender as a living thing itself. All the environmental changes that ruin the function of protein can harm an organism or even snuff out its life. The delicacy of an organism, as compared with a stone, for instance, would seem but a reflection of the delicacy of the protein that makes up the organism.

It was no surprise to biochemists, then, to find that chromosomes are largely protein in nature. They would have to be, it seemed. What else but the compound that was "of first importance" could possibly make up the chromosomes that determined the inherited characteristics of the organism?

But the chromosomes were not merely protein, after all, for it turns out that not all protein is "merely" protein. Some proteins are indeed all protein, in the sense that no portion of their substance differs markedly in characteristics from any other portion. The protein in egg white is an example of this; it is a *simple protein.*

On the other hand, *hemoglobin,* the blood protein that carries oxygen from the lungs to the body generally, is not a simple protein: it can be split into two substances, *heme* and *globin.* While the latter is a simple protein, the former is not protein at all, but an iron-containing substance, with none of the properties ordinarily associated with protein. In hemoglobin, this nonprotein portion is tightly joined to the protein. Hemoglobin is, therefore, a *conjugated protein,* the word "conjugated" originating from Latin words meaning "joined with."

Other conjugated proteins have, joined to the simple protein portion of the substance, various types of carbohydrates, lipids, pigments, metals other than iron, and so on. The particular protein of chromosomes is a conjugated protein but the nonprotein portion is none of the substances I have mentioned. It is, instead, a rather curious substance, first discovered a century ago.

In 1869, a young German chemist, named Friedrich Miescher, isolated a substance from tissue, that turned out to be neither carbohydrate, lipid, nor protein. Since he had obtained it from cell nuclei, Miescher named it *nuclein.* In time, the substance turned out to have acid properties, so that it was renamed *nucleic acid.*

It was this substance that was eventually found to be joined to the protein of chromosomes; as a result the substance of chromosomes has been given the name *nucleoprotein.*

Time passed, and during the first third of the twentieth century, biochemists were busily engaged in studying viruses, entities that caused disease but were so small that they could not be seen under the microscope. In 1935, the American biochemist, Wendell M. Stanley, isolated the *tobacco-mosaic virus* (which caused a disease of the tobacco leaf) in the form of crystals.* These crystals proved to be protein in nature.

The virus was not composed of cells but was, rather, a fragment no larger, at most, than a chromosome. Like a chromosome, the virus had the faculty of replicating itself once it got within the cell. And if there was this functional resemblance, another—a chemical—resemblance was soon to be discovered.

It turned out that the tobacco-mosaic virus was more than protein only. It contained nucleic acid as well, and was therefore a nucleoprotein. Since then, many other viruses have been isolated and analyzed, and all, without exception, have proved to be nucleoproteins.

This presented biochemists, by 1940, with a clear picture. (Two types of entities were known to replicate themselves. These were the chromosomes within the cell and the invading viruses from outside the cell. And both were nucleoprotein in nature!

Reduced to chemical terms, then, the answer to the problems of genetics lay in the nature and structure of nucleoprotein.

Variety

To the chemists of 1940 and before, however, the problem of nucleoprotein was first and foremost the problem of protein. The structure of the nonprotein portions of the substance was, in their experience, comparatively simple. It was the protein portion that counted.

The proteins were not merely complex and delicate; they existed in a tremendous variety of forms. That in itself made the subject of protein structure at once fascinating and formidable.

To show you what I mean, let me try to give you some idea of this variety.

Within the body, there are thousands of chemical reactions proceeding constantly; as yet, the total number cannot be estimated. Nevertheless, consider the fact that

---

* For this, Stanley shared in the 1946 Nobel Prize in Physiology and Medicine.

all the complex substances in food must first be broken down into smaller fragments, and then the smaller fragments must be absorbed and put together into new complex substances suitable to the eater. Some of the foodstuffs absorbed must be broken down to produce energy, and the wastes left behind must be eliminated. Special substances needed by the body must be produced out of other substances in the foodstuffs—and every change seems to take place by dozens of interrelated steps.

Almost none of the individual chemical reactions that proceed so easily and smoothly in the body will proceed in a test tube, if the reacting materials are isolated and kept at body temperature. To produce these reactions, one must add something that is extracted from living (or once-living) tissue. That something is an *enzyme* (en'zime).

An enzyme is a *catalyst* (kat'uh-list)—that is, it is a substance which, in small quantity, will make a chemical reaction proceed much more quickly than it would otherwise, yet without the catalyst itself being permanently altered in the process. The enzyme does this by supplying a surface on which the substances can react with a smaller input of energy and, therefore, with much greater rapidity.

The subject is a complex one, but I will give you a simple analogy to show what I mean. A brick resting on an inclined plank will not slide down the plank despite the pull of gravity, because friction will hold it in place. In order to get it moving, it has to be shoved; that is, there must be an input of energy. Once it starts moving, it may continue sliding to the bottom, or it may grind to a halt. However, suppose both the surface of the plank and the bottom surface of the brick are covered with a thin hard layer of smooth wax. Now the brick will slide down under the pull of gravity, without any push at all, and it will slide down more quickly. The enzyme acts, so to speak, like those layers of wax.

Now, almost each of all the thousands of reactions in the body is catalyzed by a specific enzyme. Not the same enzyme, mind you, but a different one in each case. Each reaction has its own enzyme; and every enzyme is a protein, a *different* protein.

The human body is not alone in having thousands of

different enzymes—so does every other species of creature. Many of the reactions that take place in human cells also happen in the cells of other creatures. Some of the reactions, indeed, are universal, in that they take place in all cells of every type. This means that an enzyme capable of catalyzing a particular reaction may be present in the cells of wolves, octopi, moss, and bacteria, as well as in our own cells. And yet each of these enzymes, capable though it is of catalyzing one particular reaction, is characteristic of its own species. They may all be distinguished from one another.

It follows that every species of creature has thousands of enzymes and that all those enzymes may be different. Since there are over a million different species on earth, it may be possible—judging from the enzymes alone— that different proteins exist by the billions!

More Variety

The potentialities of variation in proteins can be shown in another way, too.

The human body can form *antibodies*. These are substances which react with invading microorganisms, or with the poisonous substances that they produce, to counter the effect of the microorganism or of its poison, and thus make us immune to their action. It is in this way that the body fights a disease such as measles. The antibodies formed against the measles virus then persist in our body, or else future contact with the virus stimulates their rapid production (the body having learned the recipe, so to speak), and we remain immune to measles forever after.

Again, all of us who live in cities are constantly exposed to poliomyelitis and other serious diseases. Most of us build up antibodies against them and thus possess sufficient resistance to remain unaffected. A few unfortunates, however, succumb.

Antibodies are also built up, on occasion, against essentially harmless substances that may be present in pollen, in food, or in other parts of the environment. When we are exposed to those substances, there is a reaction between them and the antibody, and this somehow produces a number of uncomfortable symptoms, such as sneezing, swelling of the linings of the nose and throat, reddening of the eyes, blotching of the skin, asthma. We then say that we are allergic to this or to that.

Such sensitivities to specific substances can be built up deliberately. A rabbit can be injected with a particular substance; it will then build up an antibody against it. Blood serum taken from the rabbit will be found to contain the antibody, which will react with the substance to which the rabbit was made sensitive, and to no other.

There seems scarcely any limit to the number of different antibodies that can be produced. Each bacterium, each bacterial toxin, each strain of virus, each protein (and some nonprotein) component of food or of anything else, will bring about the production of a particular antibody that reacts with it and with nothing else.

An antibody that works against one strain of a particular virus will not work against an even slightly different strain of the same virus. That is why we don't work up any decent immunities against such diseases as the common cold and influenza. We produce antibodies, to be sure, but the next time we are exposed, it is almost invariably to a different strain, and our antibody is useless.

It turns out that every antibody is a protein and every different antibody is a different protein. The versatility and variety of antibodies are therefore further evidence of the versatility and variety of protein.

There are in organisms proteins that are neither enzymes nor antibodies, however, and you might think that there, at least, a standard material might exist. For instance, certain proteins form important structural components of connective tissue or of muscle. The former is *collagen* (kol'uh-jen), the latter *actomyosin* (ak'toh-my'oh-sin).* There is also hemoglobin, a protein already mentioned.

Even these, however, differ from one species to another. It is possible to produce antibodies to components of

---

* The complicated names given to most chemical substances have reasons and meanings. Usually, however, to attempt to trace those meanings would take us far afield. I will give the meanings, therefore, only when they have immediate significance. Otherwise, I will ask you to take the chemical names on faith. I will, at least, give the pronunciations when difficulty might be encountered.

human blood, for instance, which will react only with human blood. (That is how old, dried blood can be identified as human blood and not chicken blood, when murder trials demand such distinctions.)

Sometimes an antibody for chicken blood will react faintly with duck blood, or an antibody for dog blood will react faintly with wolf blood. Such faint cross-reactions are evidence of the closeness of two species in their evolutionary development.

We can summarize it all by saying that each species has its characteristic proteins and enzymes; that each individual has them; that each cell has them.

The key word is enzymes, for each organism builds its proteins through a long series of reactions that are catalyzed by enzymes. If organisms differ in substances other than protein, we may be sure that those substances, too, were built up through the catalytic activity of particular enzymes.

Enzymes in Disorder

A variation in quantity of a single enzyme out of many can produce startling changes, not only in the cells making particular use of that enzyme, but in the entire organism.

Thus, there is a brown-black pigment formed by cells of the skin in a series of reactions, each controlled by a particular enzyme. If those enzymes are all present in quantity, the pigment is formed in considerable amounts and the skin is swarthy, the hair black, the eyes brown. If one of those enzymes is formed in rather small quantities, the production of the pigment is held down; the skin is fair, the hair blond, the eyes blue. Occasionally, it may happen that an individual is born with an incapacity to form one of the enzymes. In that case, no pigment is formed. Skin and hair are white, and the eyes are pink because the blood vessels become visible in the absence of pigment. Such people are albinos.

In other words, what we consider an inherited characteristic (the color of one's hair or eyes), or one of the more startling mutations (the appearance of albinism), may be brought about not merely through the activity of cells, but through the variation in quantity of but a single enzyme within those cells.

Sometimes we cannot trace the path from enzyme to final effect quite so easily. The absence of an enzyme, or the imbalance of several, may prevent a normal reaction from taking place, or perhaps will bring about a reaction that ordinarily does not take place. Some substance will not be formed that should be formed, or it will be formed in excessive amounts. In either case, this will in turn affect the workings of other enzymes, which will upset the workings of still others, and so on. Any interference with enzyme action, at almost any point, will set off a chain reaction that can end almost anywhere.

There is an enzyme called *phenylalaninase* (feh'nil-al'an-in-ays) which on rare occasions may be absent in a human being. The reaction catalyzed by the enzyme is among those producing one of the raw materials out of which the brown-black pigment (which I mentioned before) is formed. In the absence of this enzyme, it is difficult to form the pigment, and the individual is blond. But, in addition, for reasons we don't yet know, that same individual without the enzyme will be found to suffer from a condition called *phenylpyruvic oligophrenia* (feh'nil-py-roo'vik ah'li-goh-free'nee-uh), which involves serious mental retardation.

There are many cases in which the characteristics of an organism can be traced back to the enzyme balance within the cell. From all that biochemists have been able to learn, it seems reasonable to suppose that all the characteristics of an organism are but the visible expressions of the enzyme balance.

If we are to engage ourselves in working out the puzzle of inheritance, then, we come down to two questions:

1. What is there about protein that enables it to form so many different enzymes?

2. What is there about chromosomes that enables them to bring about the formation of certain particular enzymes and no others?

To answer these questions we must plunge into a sea

of chemical language, of symbols and formulae. To try to follow the fine details of genetics without doing so would be like watching a television drama without the audio portion. You would get a general idea of the action, but you would never really learn what's going on.

CHAPTER 3

The Chemical Language

Atoms

The language of chemistry begins with the *elements*.
The elements are those substances which cannot be
broken down (by the ordinary methods developed by
nineteenth-century chemists) into simpler substances.
There are, altogether, 103 elements now known. Some of
these have been made only in the laboratory and, except
for man's interference, are not known to exist on earth.
Others do exist on earth, but are quite rare. Still others,
while common indeed, are of no importance to living
tissue.

In fact, for the purposes of this book, we need concern ourselves with exactly six elements, no more:

1. *carbon*
2. *hydrogen*
3. *oxygen*
4. *nitrogen*

5. *sulfur*
6. *phosphorus*

All are quite common, and four are frequently encountered. Coal, for instance, is almost pure carbon. So is charcoal, soot and the graphite in pencils. Diamond, too, is a special form of carbon.

Again, ninety-nine percent of the air we breathe is a mixture of oxygen and nitrogen in a 1:4 ratio, while sulfur is occasionally seen as a bright-yellow solid. Hydrogen is a light, flammable gas, sometimes used to fill balloons. Phosphorus is a reddish solid.

All substances are built of tiny *atoms*. Twentieth-century science has shown that the atoms, though incredibly small, are nevertheless exceedingly complex systems of still smaller particles. For the purposes of this book, however, we do not have to worry about the internal structure of the atom—it is enough to know that the atom is an exceedingly tiny object.

Each element is made up of an atom or atoms that are different from those of all the other elements. There are, therefore, 103 different kinds of atoms known, one for each element. Since we are going to deal with only six elements, we need to worry only about six different kinds of atoms, no more. These take their names from the elements and are therefore: (1) the carbon atom; (2) the hydrogen atom; (3) the oxygen atom; (4) the nitrogen atom; (5) the sulfur atom; and (6) the phosphorus atom.

Since we will be mentioning these atoms frequently, it would be convenient to have a shorthand method of referring to them. Chemists make use of abbreviations for this purpose, and these particular six elements are referred to, through international agreement, by their initial letters.

The carbon atom is therefore referred to as C, the hydrogen atom as H, the oxygen atom as O, the nitrogen atom as N, the sulfur atom as S, and the phosphorus atom as P.

So we begin with a stroke of good fortune. In ordinary language, we have to deal with 26 different letters, each expressed in two forms—capital and lower case. Then we have nine digits for use in forming numerals, and a variety of symbols for punctuation and other purposes. (My typewriter is equipped to produce 82 different symbols, and they are too few for my needs, actually.) In the chemical language, on the other hand, we make our start with just six symbols.

Atoms do not usually exist in isolation here on earth. Almost always, one atom is found associated with one or more other atoms. When the association is among atoms all of the same kind, we have the elements I spoke of at the beginning of the chapter. Sometimes the association is among atoms of two or more varieties, and then we have a *compound* (which comes from Latin words meaning "to put together").

Any group of atoms (alike or different) forming a close association that does not fall apart of its own accord but persists at least long enough to be studied is called a *molecule,* from a Latin word meaning "a small mass."

If atoms are the letters of the chemical language, then molecules are the words. But in order to put the chemical letters together to form chemical words, we have to know something about the rules of chemical spelling. When we deal with letters of the English language, we know we are restricted in word formation. If we write a "q," we know that the next letter must be "u." If we see an "x," we know that it is not likely to be the beginning of a word. If we see a letter combination such as "zwhf," we know at once that no English word is involved.

Chemical spelling has its rules, too, and we should not be surprised if the rules are rather different from those of English spelling. To begin with, the oxygen atom (O) and the sulfur atom (S) each have two possible "places of attachment" with other atoms, like English letters in the middle of a word, which each have one letter before and one letter after. The hydrogen atom (H) has only one place of attachment, like an English letter at the beginning or end of a word.

The nitrogen atom (N), however, has three places of attachment, and the carbon atom (C) no less than four. There we depart from any similarity to letter connections in ordinary language. (The phosphorus atom represents a special case, and I will take it up later on, when it becomes important to do so.) *

We can denote the places of attachment on each atom by little lines, called *bonds,* added to the symbol of the elements. In this way, the rules of chemical spelling can be indicated as in Figure 1.

carbon atom   nitrogen atom

oxygen atom   sulfur atom   hydrogen atom

Figure 1. *Atoms and Bonds*

Molecules

It is easy to construct simple molecules out of atoms by making use of the system of bonds shown in Figure 1. The first thing we might try to do is to set hydrogen atoms at each bond of the other atoms, in the manner shown in Figure 2.

The results are the *structural formulas* of actual well-

---

* The actual rules of atom attachments are a bit more complicated than I have just stated. Under some circumstances, for instance, the carbon atom will attach at only two places. Again, the nitrogen atom has room for a fourth, and the sulfur atom for a third and fourth, place of attachment. In this book, however, we can ignore these refinements. I only write this footnote to warn you, in all honesty, that things are sometimes more complex than I say.

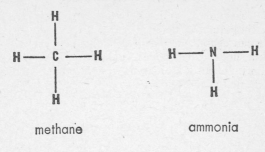

Figure 2. *Simple Molecules*

known substances. There is no need to say anything concerning *water*. As for *methane* (meth'ayn), that is a flammable gas that makes up much of the "natural gas" used for cooking and for the heating of homes. *Ammonia* is a gas with a choking odor. (The ammonia sold in the supermarket is not the substance itself, but a solution of the gas in water). *Hydrogen sulfide* is a gas with a foul rotten-egg odor often encountered in school chemistry laboratories, or emanating from stagnant bodies of water.

Chemists are so familiar with the structural formulas of these simple molecules that they usually do not bother writing them out with bonds. They merely list the different kinds of atoms and, if more than one of a particular kind is present in the molecule, they note the number. Thus, methane is written CH_4, ammonia NH_3, water H_2O, and hydrogen sulfide H_2S. When molecules are written in this fashion, we are making use of *empirical formulas*. For small molecules, the simple empirical formulas will do perfectly well.

Sometimes neighboring atoms may be held together by

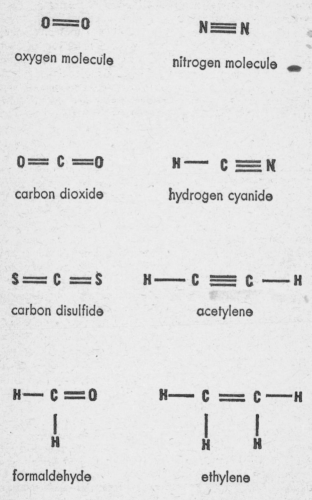

Figure 3. *Double and Triple Bonds*

the use of two bonds (a *double bond*), or even three (a *triple bond*). Some examples are given in Figure 3.

When two oxygen atoms are held together, both bonds of each atom being used, the result is a molecule made up of one kind of atom. A substance made up of such molecules is an element. The oxygen in the atmosphere is not composed of single atoms but of molecules made up of two atoms each. The oxygen of the atmosphere may therefore be referred to as *molecular oxygen*. In the same way, the nitrogen of the atmosphere is made up of two-atom molecules, with the atoms held together by all three bonds of the nitrogen atoms. Gaseous hydrogen is likewise made up of two-atom molecules, the pair of hydrogen atoms being held together, of course, by a single bond, since one bond is all that each hydrogen atom possesses.

Atoms of different types can also be held together by more than one bond, as in the case of *carbon dioxide* or *hydrogen cyanide*. However, the existence of double bonds and triple bonds introduces no change in the rules of bonding. Count the bonds attached to each atom in any of the molecules represented in Figure 3 and you will find that the oxygen and sulfur atoms always have two bonds attached, the nitrogen atom three, the carbon atom four, and the hydrogen atom one.

In writing empirical formulas, double and triple bonds are ignored. You still merely count the atoms. Thus molecular oxygen is O_2, molecular nitrogen is N_2, carbon dioxide is CO_2, hydrogen cynanide is HCN, and so on.

Carbon in Chains

So far, the molecules for which I have written the formulas are very simple ones. Continuing the analogy with words, these formulas are "words of one syllable."

The fact that more complicated molecules exist in living tissue is the result of the unique properties of the carbon atom, which is present in all living tissue. Carbon atoms have a remarkable ability to join together to form long and stable chains. Since the carbon atom has four bonds, these chains can be branched. As an example of what I mean, consider the molecule shown in Figure 4.

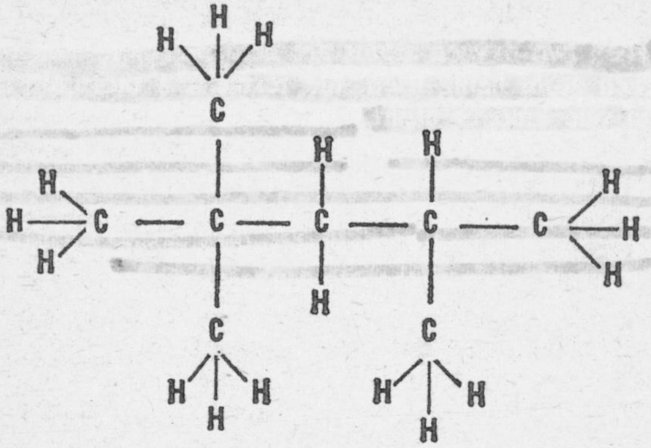

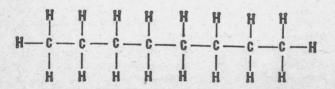

Figure 4. *Isooctane*

This molecule is known as *isooctane*. It contains eight carbon atoms arranged in a branched chain. Those carbon atom bonds that are not connected to other carbon atoms are connected to hydrogen atoms; if you count, you will find that there are eight carbon atoms and eighteen hydrogen atoms. Because its molecule contains carbon and hydrogen atoms only, isooctane is a member of a class of compounds called *hydrocarbons*. The common substance gasoline is a mixture of different hydrocarbons, with isooctane an important component.

The empirical formula of isooctane is C_8H_{18}, but once we enter the world of carbon-containing molecules, empirical formulas are no longer useful. It is possible, for instance, to arrange eight carbon atoms in a straight line, as shown in Figure 5.

Figure 5. *Normal Octane*

This represents the molecule of *normal octane,* with properties somewhat different from those of isooctane. The difference in properties means that isooctane and normal octane are in fact two distinct compounds, and yet both have the empirical formula C_8H_{18}. (And in both, you will notice, each carbon atom has four bonds, and each hydrogen atom one.)

In other words, what counts in making one molecule different from another is not merely the nature of the atoms involved and the number of each kind, but the arrangement of the various atoms. It is for this reason that, in dealing with the complex substances of living tissue, we must deal with structural formulas or we are lost.

As structural formulas grow long and complicated, it becomes convenient to be able to refer to specific portions of the molecule, to particular atomic combinations that often occur within molecules. To use the analogy with words, this would be like breaking up long words into single syllables for additional ease in pronunciation.

Thus, consider the combination of atoms shown in Figure 6.

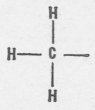

Figure 6. *The Methyl Group*

It is made up of a carbon atom with hydrogen attached to three of its bonds. The fourth bond, unoccupied in the figure, can be attached to almost any type of atom. If it were attached to a hydrogen atom, the result would be methane (as you will see, if you look back at Figure 2). For this reason, the combination of one carbon atom with three hydrogen atoms is called the *methyl group.* In the

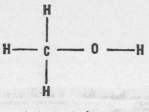

methyl alcohol

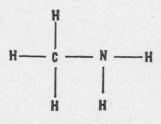

methyl amine

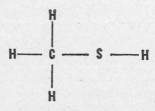

methyl mercaptan

Figure 7. Atom Groups

formula for isooctane (Figure 4), you will see five methyl groups, each of which is attached to a carbon atom.

To save space, the methyl group can be written after the fashion of empirical formulas, CH_3—. Note the dash, however, which represents the unoccupied bond. (The methyl group is not a molecule. In the type of molecules we deal with in this book, all the bonds of the various atoms are occupied. The methyl group is therefore merely the fragment of a molecule: a "syllable," so to speak, within the "word.")

The methyl group can be attached to atoms other than hydrogen or carbon. It is often attached to atoms of oxygen, nitrogen, or sulfur, and I present examples of this in Figure 7.

Each of these is what might be called a "two-syllable" molecule. The methyl group is, in each case, one syllable; what remains is the second syllable.

The oxygen-hydrogen combination in methyl alcohol can be written —OH. The name of this group is a shortened version of the names of the two atoms making it up. It is the *hydroxyl group*.

The combination of nitrogen and two hydrogen atoms present in methyl amine can be written —NH_2. One more hydrogen and it would be ammonia, and it is from that compound that the name of this *amine group* is derived. The sulfur-hydrogen combination in methyl mercaptan, —SH, is the *thiol group*, the prefix "thi-" coming from the Greek word for sulfur.

Sometimes a common atomic grouping will have two unused bonds. A carbon atom and an oxygen atom can be linked by a double bond, and the carbon atom will then still have two bonds left unoccupied. The situation can be represented as =CO. This is the *carbonyl group*, and if you will look back at Figure 3, you will find a carbonyl group present in the formula for formaldehyde.

Again, two sulfur atoms might be held together by a single bond. Each would have one bond left unoccupied, making a total of two. Such a group, —SS—, is a *disulfide group*.

One of the organic compounds known longest to man in reasonably pure form is *acetic acid* (uh-see'tik), a name derived from the Latin word for vinegar. Vinegar is in fact just a weak solution of this acid. The formula for acetic acid is given in Figure 8.

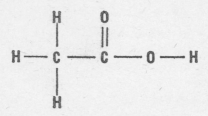

Figure 8. *Acetic Acid*

As you see, acetic acid is a "three-syllable" molecule. It contains a methyl group attached to a carbonyl group which is, in turn, attached to a hydroxyl group. The carbonyl-hydroxyl combination occurs very frequently in compounds, so that it is usually treated as a "syllable" in itself. The phrase "*carb*onyl-hydr*oxyl*" is shortened to the italicized portions shown, and the group is called the *carboxyl group* (kahr-bok'sil). Because the presence of a carboxyl group in a molecule tends to give that molecule the properties of an acid, it is sometimes called the *carboxylic acid group* (kahr-bok-sil'ik).

The carboxyl group is often written, for short, as —COOH. This isn't really a good representation because it makes it look as though the two oxygen atoms are attached to each other, and they aren't. I myself would prefer to write it —(CO)OH or —CO(OH), but I am quite certain that I shall never succeed in changing a century-old chemical habit.

If, for the hydroxyl portion of the carboxyl group, an amine group is substituted, the result is —CONH$_2$. This is an *amide group* (am'ide).

There are many additional groupings with which the professional chemist must deal in considering organic compounds, but we can get by with the eight I have mentioned. For convenience's sake, I shall list them here:

—CH₃	methyl group
—OH	hydroxyl group
—NH₂	amine group
—SH	thiol group
=CO	carbonyl group
—SS—	disulfide group
—COOH	carboxyl group
—CONH₂	amide group

Carbon in Rings

We are not quite through. There is still one further refinement to take into account.

Carbon atoms have a tendency to form rings. These rings make up unusually stable combinations, partic-

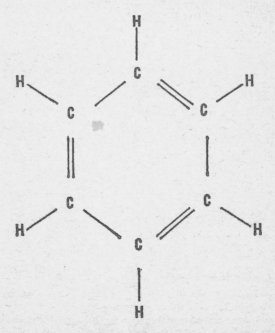

Figure 9. Benzene

ularly when composed of either five or six atoms, and most particularly when double bonds are included in alternation with single bonds. The best example is given in Figure 9.

The molecule shown is that of *benzene* (ben-zeen'). It has at its core a ring of six carbon atoms, each of which is connected to one neighboring carbon atom by a single bond and to another by a double bond. Each carbon atom also has a fourth bond, which is attached to a hydrogen atom.

The ring of six carbon atoms with its alternate-double-bond, single-bond arrangement is called the *benzene ring*. It is so stable that it is found to be part of many thousands of compounds.

Chemists have had to make use of this ring so often in writing their formulas that they have naturally sought some brief way to indicate it, and the solution most frequently used is that of presenting it geometrically. The benzene ring is written as a simple hexagon, with only the single bonds and double bonds showing, as in Figure 10.

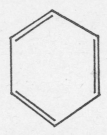

Figure 10. *The Benzene Ring*

In order to reconvert this geometrical version of the benzene ring into the molecule of benzene with each atom showing, it is necessary only to place a C at each vertex of the hexagon and to remember that all the bonds left over are attached to hydrogen atoms. This becomes such second nature to chemists that complex ring systems are understood at a glance.

But what if the bonds left over are attached to atoms other than hydrogen? In that case, those connected atoms

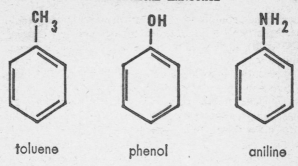

Figure 11. *Compounds Containing a Benzene Ring*

or groups of atoms are shown. I give you examples of that in Figure 11, where *toluene* (tol'yoo-een) has a methyl group attached to the benzene ring, *phenol* (fee'nole) a hydroxyl group, and *aniline* (an'ih-lin) an amine group. For greater simplicity, the added groups are almost invariably written as empirical formulas; later on

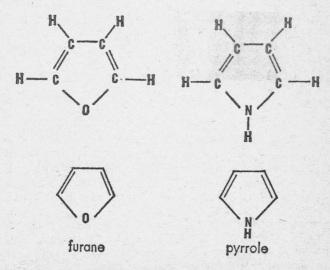

Figure 12. *Five-Atom Rings*

in the book, I shall introduce a still further simplification.

Sometimes a ring of atoms is not made up entirely of carbon atoms. Other atoms, usually nitrogen or oxygen, may be involved. In that case, the geometric representation must specify the atom other than carbon. Only then can you be certain that, at any angle of a figure where no specific atom is shown, the atom that belongs there is carbon. As examples, two compounds are written out in full and in geometric form in Figure 12.

In the case of these two compounds, *furane* (fyoo'rane) and *pyrrole* (pih-role'), only five atoms are involved in the ring, so that the geometric representation is that of a pentagon.

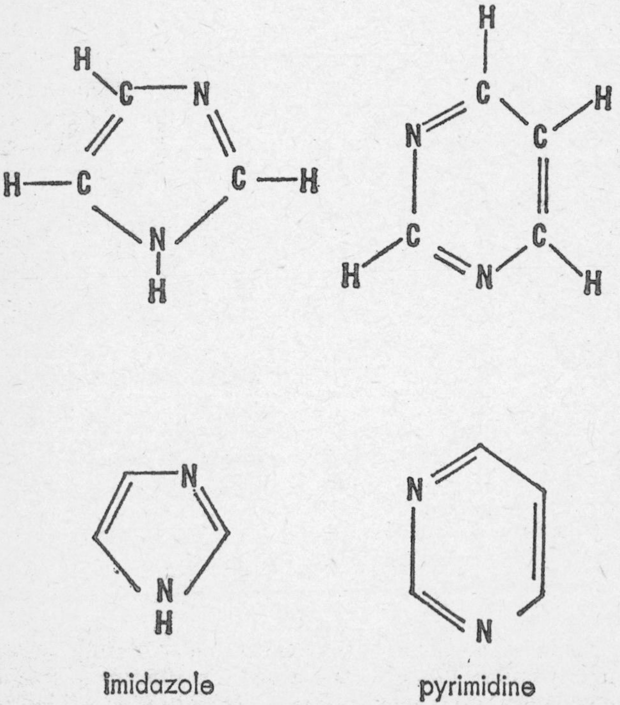

imidazole   pyrimidine

Figure 13. *Two-Nitrogen Rings*

It is also possible for six-membered rings to contain atoms other than carbon, of course, and it is even possible for them to contain more than one such atom. Some examples are shown in Figure 13. *Imidazole* (im′id-az′ole) is a five-membered ring with two nitrogen atoms, and *pyrimidine* (pih-rim′ih-deen) a six-membered ring with two nitrogen atoms.

It is possible, also, for carbon atoms (with or without a few atoms other than carbon) to arrange themselves in combinations of rings. For instance, a benzene ring and a pyrrole ring can combine to form *indole* (in′dole), while a pyrimidine ring and an imidazole ring can combine to form *purine* (pyoo′reen). These are shown in Figure 14.

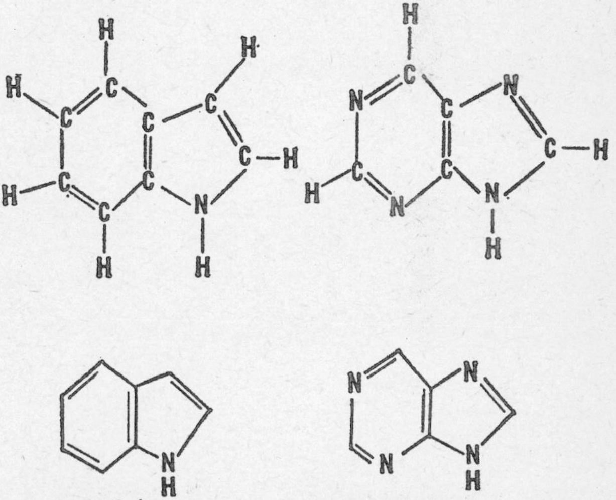

Figure 14. *Ring Combinations*

This by no means exhausts the different rings which it is possible to form, and which are found in organic compounds. As a matter of fact, chemists have sometimes prepared rather sizable booklets entirely filled with lists of the different rings and ring combinations they are likely to run across, and giving a name for each.

For our purposes in this book, however, we will need only the seven rings and ring combinations I have mentioned. For the sake of convenience, I will now list all of them together, geometric representations only, in Figure 15.

The 8 groups and 7 rings listed in this chapter give us just about all the basic "syllables" we will need in making

benzene ring   furane ring   pyrrole ring

imidazole ring   pyrimidine ring

indole ring   purine ring

Figure 15. *The List of Rings*

use of the chemical language. (I will add one or two additional items as we proceed.)

This may strike you as, if anything, a bit too simple. Will such a limited "syllabary" help us to explain the vast complexity and variety of protein?

Oddly enough, it will, as we shall soon see.

CHAPTER 4

The Building Blocks of Proteins

Giant Molecules

At the beginning of the nineteenth century, when chemists first became aware of the existence of atoms, the first molecules with which they dealt were small ones, the "one-syllable words" mentioned at the beginning of the previous chapter. It was impossible, however, to deal with organic substances without stumbling onto truly giant molecules.

Fortunately, these giant molecules turned out in every case to be large only in that they consisted of a number of small molecules joined together like beads on a string. It proved possible to treat the large molecule in such a way as to free the small-molecule units from their connections with neighboring units. This is usually done by heating the large molecule in acid solution.

Whereas the large molecule, while intact, is very difficult to study, the small units, once they are independent, are easily handled. From the knowledge gained of the structure of these building blocks, it was often possible to deduce the structure of the giant molecule in its intact state.

If we consider the small units "words" and the large molecule a "sentence," the problem is similar to that of a man facing an inscription in a foreign language of which he only has a smattering of knowledge. If he were to read an entire sentence at a breath, he might not be able to follow it. However, if he were to read it a word at a time, looking in the dictionary for words he didn't remember the meaning of, he could very likely puzzle out the entire sentence.

The first large molecule, or *macromolecule,* studied in this fashion proved unexpectedly simple. As early as 1814, it was discovered that starch, if heated long enough in acid solution, broke down into units all of identical structure. The structure was *glucose,* a type of sugar with a molecule only about half the size of that of ordinary table sugar. Its empirical formula is $C_6H_{12}O_6$, showing that the molecule contains only twenty-four atoms. However, hundreds and even thousands of these units would string together to form a single starch molecule, which was thus made up of hundreds of thousands of atoms.

The stiffening substance in wood, *cellulose,* was also found to break down into glucose, the same glucose that was to be found in starch. In cellulose, however, the glucose units were hooked together in a fashion somewhat different from that found in starch.

As time went on, other macromolecules were also found to be built up of long chains of a single unit. Rubber is a good example, being built up of molecules of *isoprene* (igh'soh-preen), a rather simple five-carbon hydrocarbon.

In the twentieth century, chemists learned to build macromolecules that do not occur in nature. They devised methods of stringing together many molecules of one particular unit or another (sometimes a mixture of two units), so as to produce artificial rubbers, synthetic fibers, and a large variety of plastics.

All these macromolecules, natural and synthetic, were alike in being large, and in being built up of thousands of units per molecule. Yet, although large, they somehow lacked complexity. You will see what I mean if you consider that a long string of beads, all of them identical in size and color, is certainly not complex. There is no scope for creativity in stringing such beads: one string may be longer than another, or may be arranged in a double loop instead of a single one, but they are all much alike.

Size has its uses, of course. The glucose units lined up by the thousands to form cellulose produce a stiff, tough substance that enables a tree to stand firm in the teeth of a gale, and out of which we do not scorn to build our own shelters. Again, the starch macromolecule is an excellent way of storing the energy content of the glucose molecule in a stable, insoluble fashion until it is called for. Then the starch molecules can easily be broken down, and the individual glucose units fed into the bloodstream.

Still, macromolecules such as starch and cellulose do not play a truly active role in the life process. They are passive materials that do not act but are acted upon.

It is different with protein. Here we have a macromolecule which, like starch or cellulose, is large; like them, it is also built up of small units strung together like beads on a string. In protein molecules, however, complexity has been added to mere size. It now remains to show how this is done.

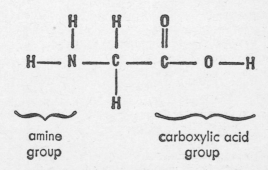

amine
group

carboxylic acid
group

Figure 16. *Glycine*

Amino Acids

About 1820, a French chemist, H. Braconnot, heated the protein gelatin in acid and obtained crystals of a sweet-tasting compound. Eventually, this was given the name *glycine* (gly'seen), from the Greek word meaning "sweet."

The structure of the glycine molecule was worked out, and it proved to be simple. It was made up of only 10 atoms, less than half the number in glucose. The formula of glycine is shown in Figure 16.

As you see, the molecule consists of a central carbon atom, which is attached by one bond to an amine group * and by a second bond to a carboxylic acid group. The remaining two bonds are occupied by hydrogen atoms. Now, a compound which contains both an amine group and a carboxylic acid group could very naturally be called an *amino acid,* and it is. Glycine is, in fact, a particularly simple example of an amino acid.

If matters had ended there, then protein, as a macromolecule, would be thought no more complex than starch or any other macromolecule. However, Braconnot went further and obtained a second amino acid out of the breakdown products of protein; he called it *leucine* (lyoo'seen), (from the Greek word for "white," because the crystals he obtained were white).

As the decades passed, additional amino acids were found by other investigators. As late as 1935, a new and important amino acid, the existence of which had not previously been suspected, was discovered among the breakdown products of protein molecules. It is these

* I hope it is clear that a chemical group can be written either from left to right or from right to left. Thus, a hydroxyl group can be written HO— or —OH; an amine group NH_2— or —H_2N; and a carboxyl group —COOH or HOOC—. In all these cases, it just depends on whether you are viewing the group from in front or from behind. The nature of the group isn't changed by this change in the observer's point of view. In the formula for glycine in Figure 16, I write the amine group "backward," as compared with the way in which it is written in the formula for methylamine in Figure 7, but this does not matter. Similarly, if the benzene ring, for instance, is viewed from behind, the double bonds seem to run counterclockwise rather than clockwise. That doesn't matter, either.

amino acids, then, that are the building blocks out of which the protein molecules are constructed.

The number of different amino acids found in living tissue is quite large. Some of them, however, are not found in protein molecules, although they occur elsewhere. Others are indeed found in protein molecules, but only in one or two rather unusual instances.

If we restrict ourselves to the amino acids found in every, or almost every, protein molecule, the number is still sizable—21. Add to this another amino acid found principally in only one protein molecule (but a very important one), and the total is 22.

Here is one way in which the protein molecule is unique. No other macromolecule, either natural or man-made, is constructed of so many different units, or even of a quarter as many.

The importance of this is plain if we go back to our string of beads. Imagine that instead of but one set of beads, all identical, you were supplied with twenty-two sets, each differing from all the others in color, shape, or size. It would now be possible to produce a variety of fascinating patterns, unexpected symmetries, pleasant gradations that would otherwise have been impossible.

And so it is with the protein molecule.

But let us take a closer look at the amino acids and see just how these differences among them appear, and in what way they lend a pattern to the protein molecule and create the possibility of virtually endless variety.

To do this as clearly as possible, I would like to introduce a schematic way of dealing with structural formulas. I would like to extend the geometric principle, by which rings of atoms are represented, to atoms that are not part of rings. (This is further than professional chemists generally go in simplifying formulas, but that doesn't matter. This book is not being written for the professional chemist. Its only purpose is to explain the chemical basis of inheritance in the simplest and most direct fashion I can think of, and if that means a little innovation, why—full speed ahead!)

In forming the geometrical figures shown in Figure 15, I explained that a carbon atom exists at every unoccupied angle. Again, any carbon bond which is not shown is assumed to be connected to a hydrogen atom.

Now let us extend this by writing a zigzag line for atoms that do not form a ring. We can continue to assume that a carbon atom is present at every unoccupied angle (and at every unoccupied end). Furthermore, we can extend the "don't-show-hydrogen" rule to atoms other than carbon.

As an example, I present the "zigzag formula" of glycine in Figure 17, and you can compare it with the formula given in full in Figure 16.

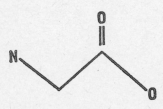

Figure 17. *Glycine (zigzag)*

The next step is to see how the other amino acids out of which the protein molecule is built differ from glycine. In general, one can say that all of them possess a central carbon atom, to which an amine group is attached by one bond, and a carboxylic acid group by another.

The differences among them arise as follows: In glycine the third and fourth bonds of the central carbon atom are both attached to hydrogen atoms. In the other amino acids, the third bond is indeed attached to a hydrogen atom, but the fourth bond is attached to a carbon atom which is, in turn, part of a more or less complicated group of atoms called a *side chain*.

The difference can be made perfectly clear if you look at Figure 18, which shows the general amino acid formula in zigzag fashion, and compare it with the zigzag formula for glycine in Figure 17.

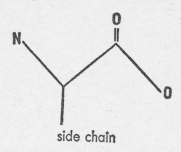

side chain

Figure 18. *Amino Acid (zigzag)*

Each different amino acid has its own characteristic side chain, and it is in the nature of this side chain that the crucial difference between amino acids is to be found.

Side Chains

Let us now look at each of the twenty-one amino acids other than glycine and consider the side chains in order to get a notion of the differences that exist. To begin with, I will present each side chain in full, with all the atoms showing—for the record.

First, there are four amino acids in which the side chain is a hydrocarbon group. One is leucine, already mentioned. The other three are *alanine* (al'uh-neen), *valine* (val'een), and *isoleucine* (igh'soh-lyoo'seen). Their side chains are shown in Figure 19.

There are two amino acids with hydroxyl groups in the side chain. These are *serine* (ser'een) and *threonine* (three'uh-neen), and their side chains are shown in Figure 20. It was threonine that, in 1935, was the last of the

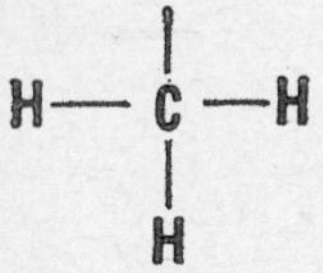

alanine side chain

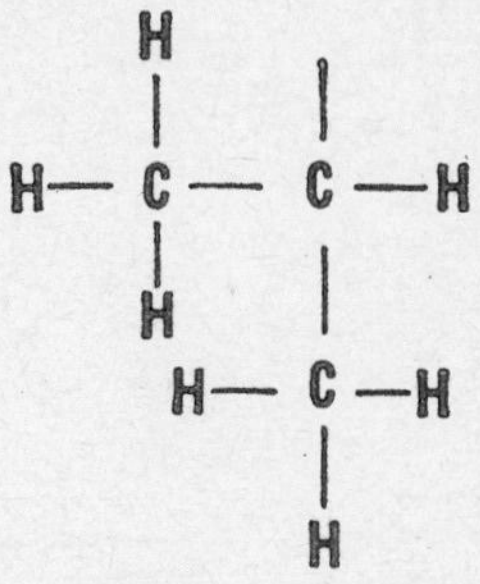

valine side chain

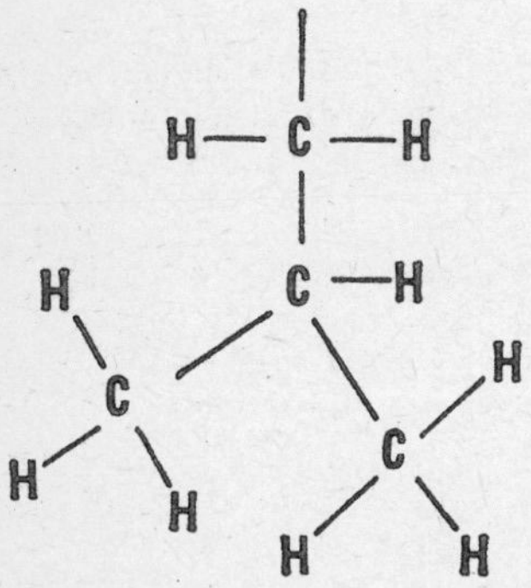

leucine side chain

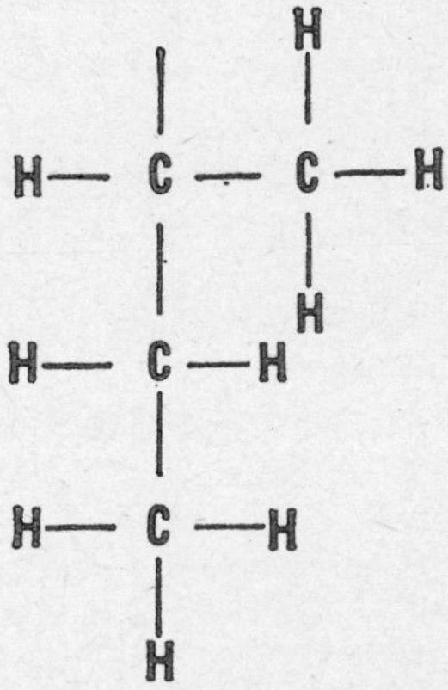

isoleucine side chain

Figure 19. *Hydrocarbon Side-Chains*

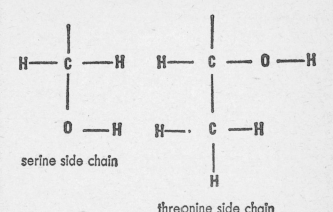

serine side chain

threonine side chain

Figure 20. Hydroxyl-Containing Side-Chains

group to be discovered. Chemists are almost certain that no other important amino acid (none, that is, which occurs in all or almost all proteins) remains to be discovered in the future.

Two amino acids contain carboxylic acid groups in the side chain. These are *aspartic acid* (uh-spahr'tik) and *glutamic acid* (gloo-tam'ik). Closely resembling these in name as well as in structure are two amino acids with an amide group in place of the carboxyl group. These are *asparagine* (uh-spar'uh-jeen) and *glutamine* (gloo'tuh-meen). All four side chains are shown in Figure 21.

Two amino acids contain amine groups in the side chain. One is *lysine* (ly'seen), and the other is *arginine* (ahr'jih-neen),* and those side chains are shown in Figure 22.

* The combination of three nitrogen atoms attached to a central carbon atom, which you see in the arginine side chain, is a *guanido group.* This is not particularly important for purposes of this book, but I mention it to remind you that there are indeed more atom groups than those mentioned in the previous chapter.

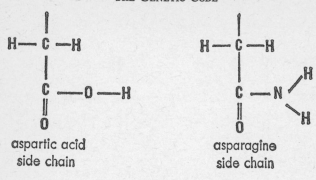

aspartic acid
side chain

asparagine
side chain

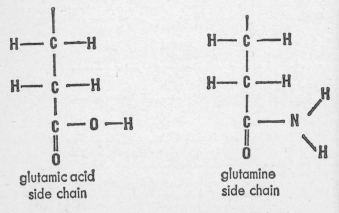

glutamic acid
side chain

glutamine
side chain

Figure 21. Carboxyl- and Amide-Containing Side-Chains

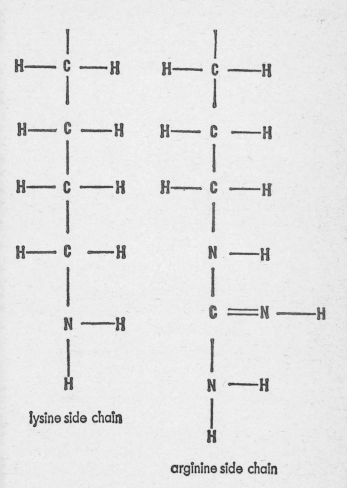

lysine side chain

arginine side chain

Figure 22. Amine-Containing Side-Chains

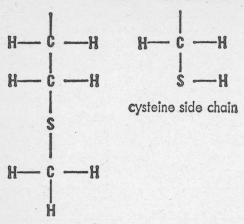

methionine side chain

cysteine side chain

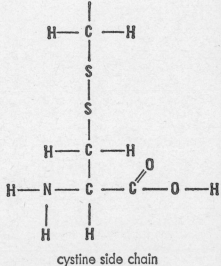

cystine side chain

Figure 23. Sulfur-Containing Side-Chains

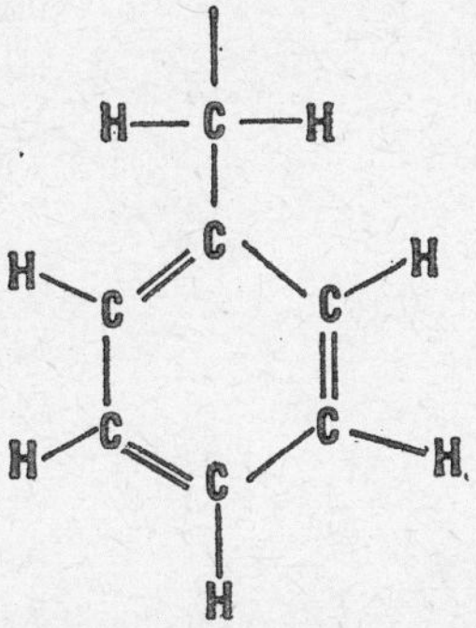

phenylalanine side chain

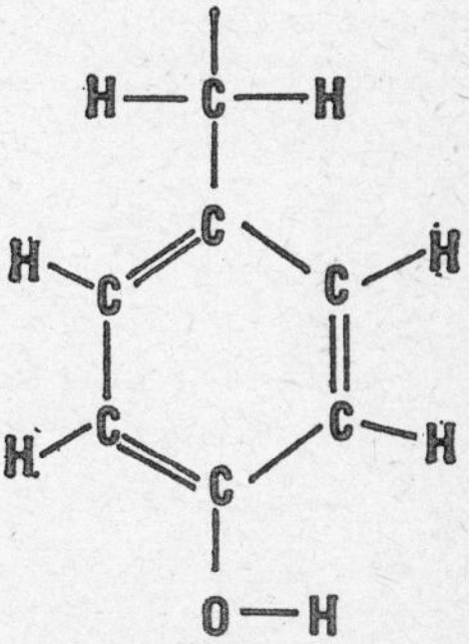

tyrosine side chain

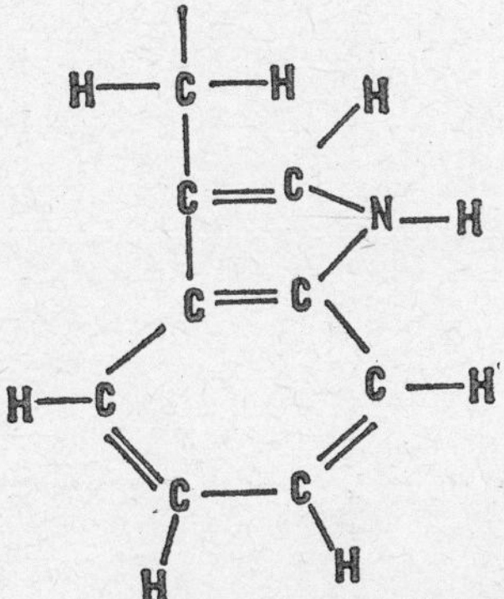

tryptophan side chain

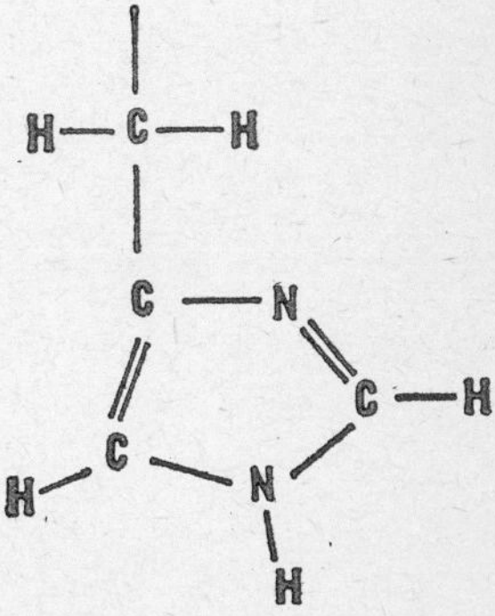

histidine side chain

Figure 24. *Ring-Containing Side-Chains*

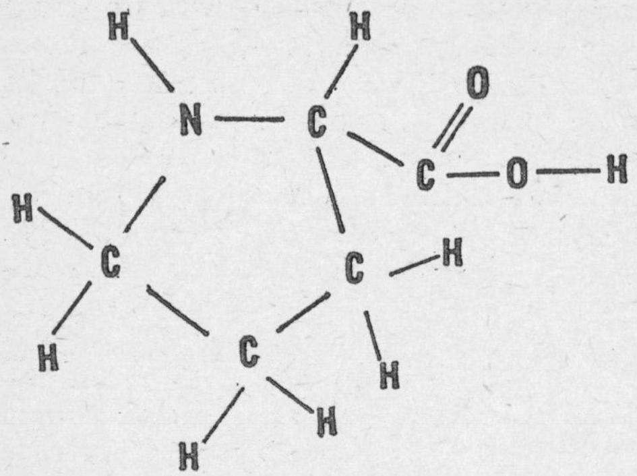

proline

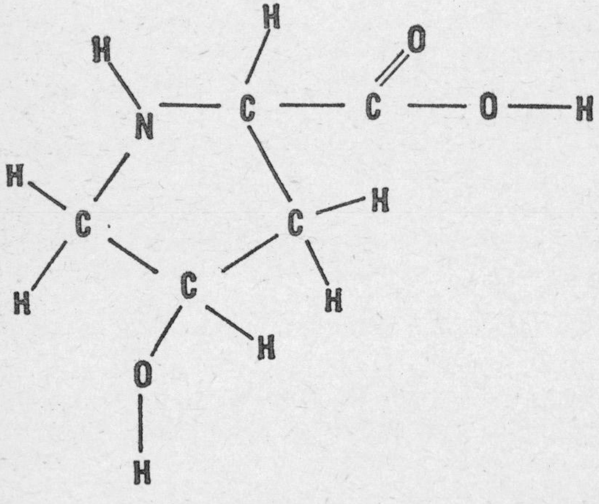

hydroxyproline

Figure 25. Proline and Hydroxyproline

Three amino acids contain sulfur atoms in the side chain. One is *methionine* (meh-thy'oh-neen), which contains a single sulfur atom between two carbon atoms (a combination sometimes referred to as a "thio-ether"). Another, *cysteine* (sis'tuh-een), possesses a thiol group, while a third, *cystine* (sis'teen), has a disulfide group. All three side chains are shown in Figure 23.

Notice that in the side chain of the cystine molecule there is an amino-acid arrangement at the end. If the molecule were written out in full, it would look like two cysteine molecules jammed together at the disulfide group. The cystine molecule is easily taken apart into two cysteine molecules, while two cysteine molecules can as easily be put together to form one cystine molecule. (This accounts for the similarity in names which is, as a matter of fact, quite a bother. If the letter "e" is omitted or inserted by accident, or if a speaker isn't careful in his pronunciation, the two amino acids are confused.)

No less than four amino acids have rings in their side chains. Two *phenylalanine* (fen'il-al'uh-neen) and *tyrosine* (ty'roh-seen), have benzene rings; one, *tryptophan* (trip'toh-fan), has an indole ring, and the fourth, *histidine* (his'tih-deen), has an imidazole ring. The side chains are shown in Figure 24.

Finally, there are two amino acids in which the side chain does a peculiar thing. It turns back on itself and joins with the amine group attached to the central carbon atom. Because of this, the formulas of the two amino acids *proline* (proh'leen) and *hydroxyproline* (hy-drok'see-proh'leen) are shown in full in Figure 25. Notice that each compound is converted into a pyrrole ring arrangement (without the double bonds) by this odd formation of the side chain. (The word "proline" is derived from "pyrrole," as a matter of fact.)

It is hydroxyproline that occurs in only one protein, by the way. That protein is *collagen* (kol'uh-jen) which makes up a good part of the connective tissue of the animal body, including, of course, our own. It is found in skin, in cartilage, in ligaments and tendons, in bone, hooves, and horns. When boiled strenuously, collagen is broken down into a familiar protein, gelatin, so that hydroxyproline appears there, too.

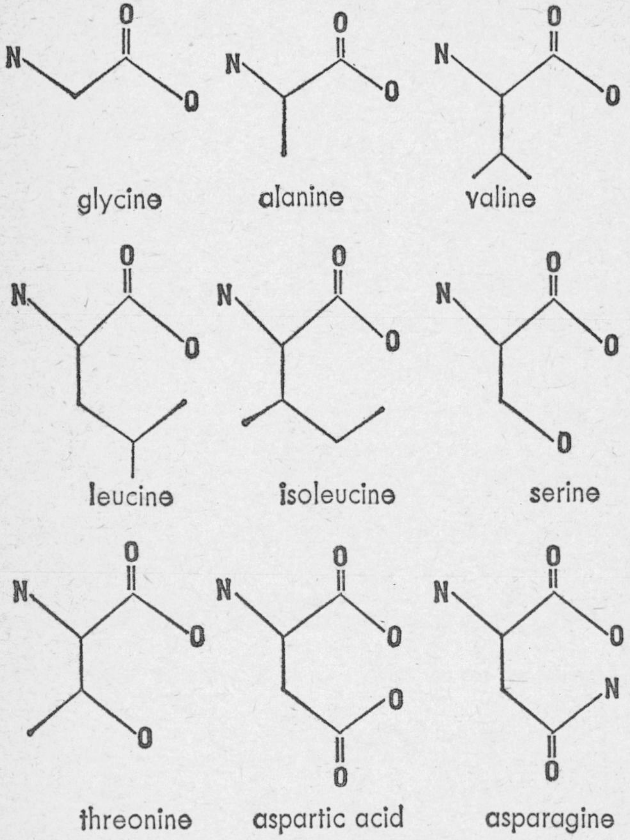

Figure 26. *The Twenty-Two Amino Acids (zigzag)*

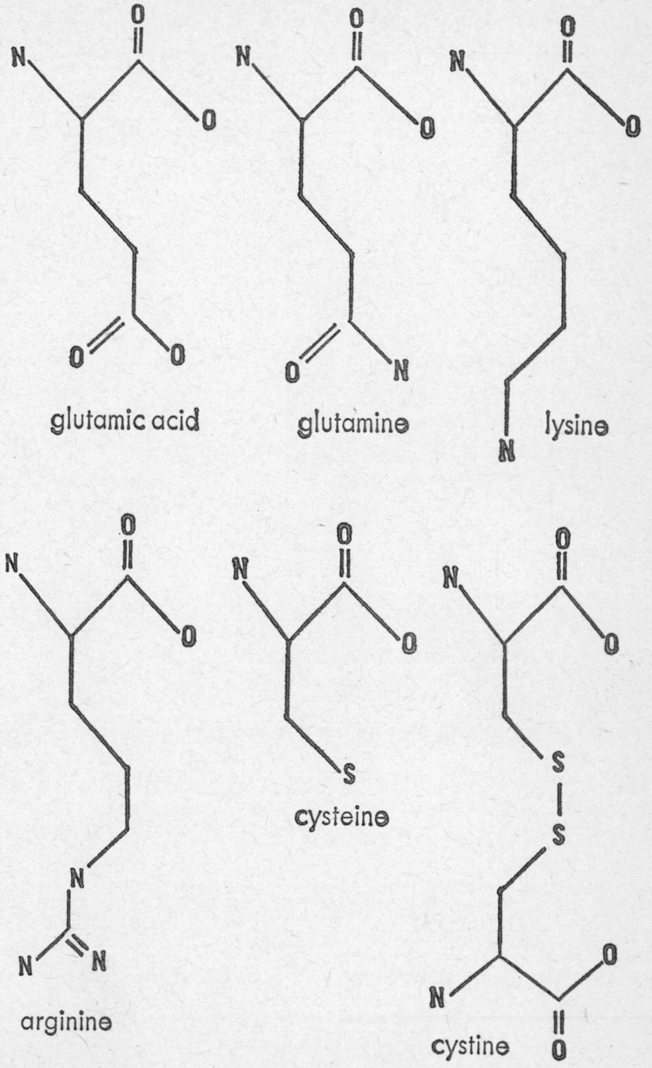

glutamic acid

glutamine

lysine

arginine

cysteine

cystine

Figure 26 (continued)

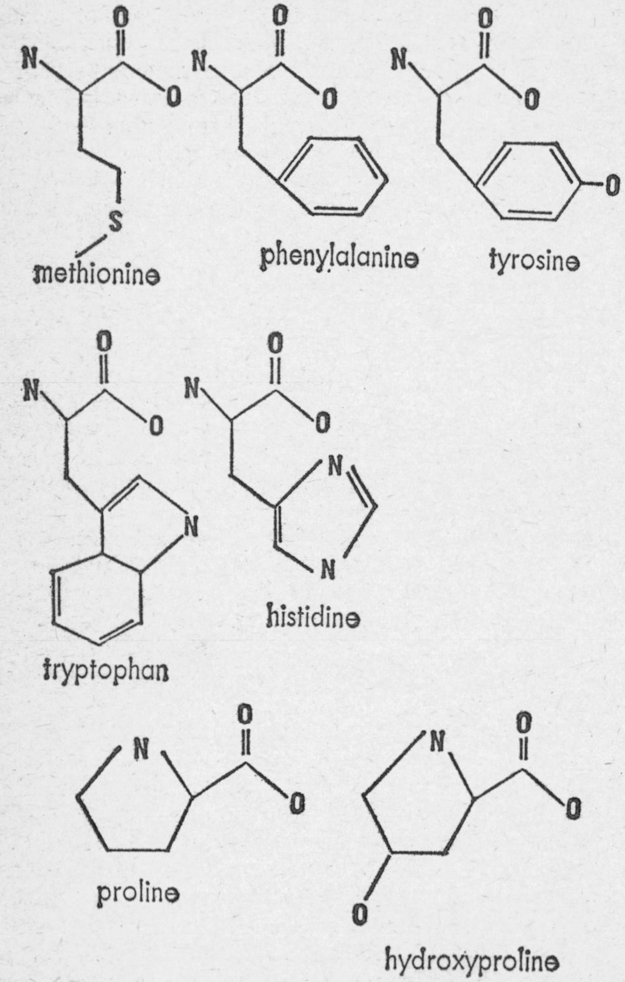

methionine

phenylalanine

tyrosine

tryptophan

histidine

proline

hydroxyproline

Figure 26 (concluded)

The roll call is complete, and we now have the twenty-two amino acids, the twenty-two "words" out of which the "sentence" of the protein molecule is constructed.* At this point I would like to summarize matters by listing all the amino acids in zigzag formula, as shown in Figure 26. I think the pattern shown in that figure makes the structural differences clear and also emphasizes the family relationships. By following the rules outlined a few pages ago, you can convert each one of those zigzags into a full formula . . . if you feel moved to do so.

Words into Sentences

Now that the "words" are in our possession, let's consider how they might be put together to form a "sentence." This problem was not solved until the first decade of the twentieth century, when the German chemist Emil Fischer made the first satisfactory demonstration of how this might be done.

He showed that two amino acids combined by the joining of the carboxylic acid group of one with the amine group of the other and that, in the process, a water molecule is lost. If we use two glycine molecules, so as to present matters at their simplest, this sort of joining takes place, as shown in Figure 27, in which the position of each atom is shown.

As you see, the hydroxyl group that makes up part of the carboxylic acid group combines with one of the hydrogen atoms of the amine group. Together, the hydroxyl group and the hydrogen atom make up a water molecule, H_2O, which is removed. By the removal of the hydroxyl group and the hydrogen atom, each glycine molecule has one bond freed, and these two join to form *glycylglycine* (gly'sil-gly'seen).

---

* I might say here that the figure twenty-two must be taken as rather arbitrary. Some biochemists consider asparagine and glutamine as mere varieties of aspartic acid and glutamic acid, and for them there are only twenty different amino acids. Other biochemists are reluctant to count hydroxyproline, which doesn't occur in proteins other than collagen, and still others tend to count cysteine and cystine as two varieties of a single structure, so that the number of different amino acids might be taken to be as few as eighteen. However, I prefer the liberal view and shall stick to twenty-two.

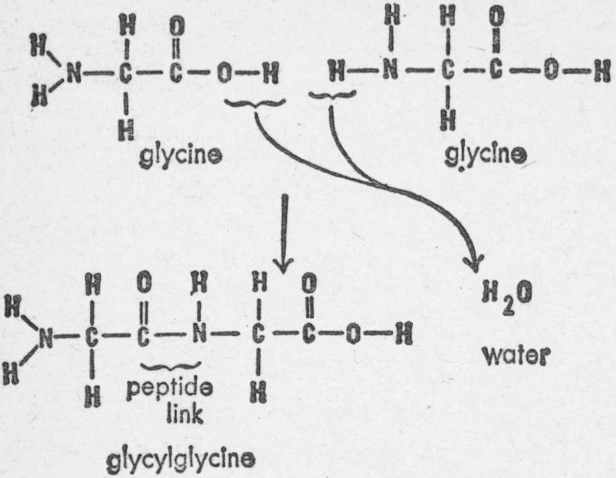

Figure 27. An Amino Acid Combination

Such combinations of amino acids are called *peptides,* from the Greek word for "digest," because they were first obtained from partially digested protein. The combination of atoms that makes up the union between the amino acids is —CONH— (in which you can see the remaining portions of the original carboxylic acid group and amine group), and this is called the *peptide linkage.*

Glycylglycine is a peptide made up of two amino acids; it is called, therefore, a *dipeptide.* (It is a "two-word sentence," so to speak.) Glycylglycine still has an amine group at one end and a carboxylic acid group at the other end, however, so that it can combine with still other amino acids at either end or at both ends. In this way, *tripeptides, tetrapeptides, pentapeptides,* and so on can be built up.*

* The prefixes "di-," "tri-," "tetra-," and "penta-" are taken from the Greek prefixes for "two," "three," "four," and "five." These numerals are frequently used in chemical nomenclature. The compound normal octane, shown in Figure 5, has eight carbon atoms, and the prefix "oct-" is from the Greek word for "eight."

There is no limit to the number of amino acids that can be put together by peptide links. A peptide made up of an undetermined number of amino acids is called a *polypeptide,* the prefix "poly-" coming from the Greek word for "many."

Suppose, then, we picture a large number of glycine molecules joined together by peptide links. The result, presented in zigzag, is shown in Figure 28.

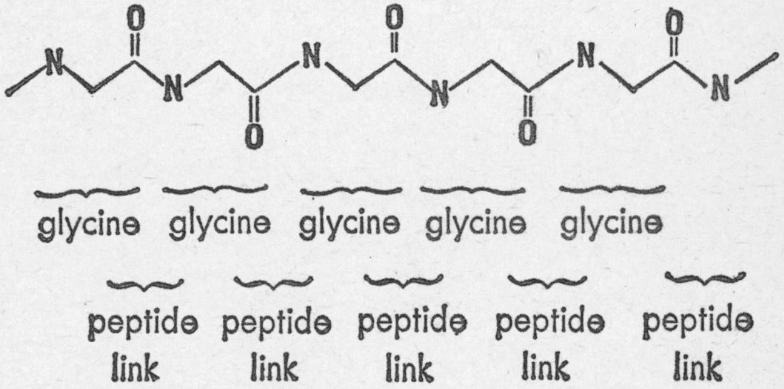

Figure 28. *Polyglycine*

This particular polypeptide, made up of glycine units only, is *polyglycine*. A polyglycine molecule is no more complex and no more capable of proteinlike versatility than any other macromolecule made up of only one or two varieties of unit. A naturally occurring polypeptide that is mostly glycine and alanine is silk, and there the lack of complexity is evident. Silk is used by those organisms that form it only in ways that take advantage of its strength as a fiber. It is little more than a kind of animal version of cellulose.

Another example is the artificial fiber, Nylon. This is made up of two units, one a dicarboxylic acid (a carbon chain with a carboxylic acid group at either end) and the other a diamine (a carbon chain with an amine group on either end). The units are held together by peptide links, and Nylon, too, is valued chiefly for its strength.

For versatility we must turn to the fact that a polypeptide chain, as it occurs in nature, is almost always made up of as many as 22 different units. Such a polypeptide chain would differ from polyglycine in that side chains would be present at periodic intervals. As you can see in the zigzag presentation of such a polypeptide chain in Figure 29, the side chains (symbolized by X) branch

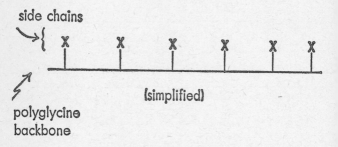

(zigzag)

side chains

(simplified)

polyglycine backbone

Figure 29. *The Polypeptide Chain*

off, alternately, in opposite directions.

A polypeptide chain, then, consists of two portions: (1) a polyglycine backbone that runs the length of the chain; and (2) a variety of side chains that emerge from that backbone.

Since we are interested only in those features of the protein molecule that lend it versatility, we will omit the common feature and concentrate on the side chains. The details of the polyglycine backbone (now that we know them) are unimportant and can be represented quite comfortably, for our purposes, by a simple straight line. The side chains can then be shown as all issuing on the same side, in a further move toward simplification. Figure 29, which shows the polypeptide chain in zigzag fashion, also includes, for comparison, this further simplification.

A protein molecule often consists of nothing more than a single polypeptide chain. Sometimes, however, it is made up of two or more polypeptide chains, bound together by cystine molecules. If you look back at the formula for cystine in Figure 23, you will see that there is an amino-acid combination at both ends. This means that one amino-acid end can form part of one polypeptide chain and the other part of a second polypeptide chain, as you can see in the simplified formula in Figure 30. The polypeptide chains are then held together by a disulfide link.

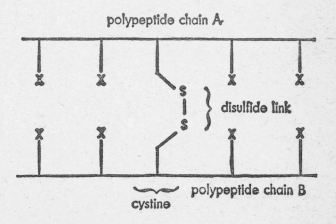

Figure 30. *Polypeptide Chains in Combination*

This disulfide link is easily broken by chemical treatments that leave the polypeptide chains themselves intact, so that chemists are able to study the chains individually. Once Fischer had established the nature of the polyglycine backbone and settled that part of the problem, chemists, in these studies, turned to the matter of the side-chain pattern, and it is to that which we, too, will now turn.

CHAPTER 5

The Pattern of Protein

Number and Order

Side chains present a varied spectrum of properties. Some, like those of tyrosine and tryptophan, are large and bulky, while others, like those of alanine and serine, are small. Some side chains, like that of threonine, carry a hydroxyl group, others do not; some like those of aspartic acid and glutamic acid, ordinarily carry a negative electric charge, others, like those of lysine and arginine, carry a positive electric charge. Most carry no electric charge.

The result is that a particular protein molecule can bristle with a pattern of side chains that may concentrate bulkiness here and not there, that may distribute negative electric charges here and positive ones there and none at all yonder.

One can picture, from this standpoint, how an antibody might work. A protein could be constructed with a side-chain pattern which would just fit the side-chain pattern on a foreign protein, or on a virus or on a key spot on a bacterial surface. The fit may be one in which a negative electric charge on the antibody meets a positive one on the invading molecule, with mutual attraction; or a bulky collection of atoms on one molecule may just fit a recess on the other. In either case, antibody and prey join tightly, and the combination then becomes harmless to the body. Of course, a particular antibody with a pattern just suited to one particular sort of molecule will fit no other (or at any rate will only fit others that are extremely similar to the one for which it is suited).

One can also picture how an enzyme might work. A particular enzyme could have a pattern of side chains such that two reacting chemicals will just fit conveniently into adjoining niches. Brought together by a middleman, so to speak, they will react with each other and leave, vacating the spot for another set of reactants, so that the reaction as a whole will then proceed at a far more rapid rate than it would if the enzyme were absent. Naturally, an enzyme just made to fit one set of reactants will not fit another.

To understand the workings of a protein, then, it becomes necessary to understand its side-chain pattern in full. This is not to say that complete knowledge of the pattern would answer all questions; it very probably would not. But without knowledge of the pattern it is quite certain that the questions would not be answered. Tracing the pattern, then, is at least a necessary step toward the answer.

The pattern can be attacked in three stages. Since I have been using the analogy of molecular structure and ordinary language in the preceding chapters, I will make use of it now to explain those three stages.

The first stage is that of learning which amino acid units are present in a particular protein molecule. This is equivalent to determining the exact words in an English sentence. The meaning of a sentence can obviously be changed if we alter even a single word. Thus, in two sentences—

John only punched Jim in his eye.
John only punched Jim in his dreams.

—the change of a key word makes a world of difference.

Once each amino-acid unit is known, the second stage consists in determining the exact order of those units along the polypeptide chain.

This would be equivalent to determining the exact word order in a sentence. The meaning of a sentence can be changed quite markedly without changing one word in it—if the order of words is rearranged. Consider:

John only punched Jim in his eye.
John punched Jim in his only eye.
Only John punched Jim in his eye.
John punched only Jim in his eye.
John punched Jim only in his eye.
and
Jim only punched John in his eye. . . . etc.

Finally, there is a third change that will require a short explanation.

The polypeptide chain is capable of a limited degree of bending. The chain is held in this bent position by weak electrical forces that exist wherever a hydrogen atom lies between two closely spaced nitrogen atoms, or between two oxygen atoms, or between an oxygen and a nitrogen atom. Such a bond is called a *hydrogen bond,* because of the central role played by the hydrogen atom.

So long as the hydrogen bonds remain intact, the polypeptide chain retains its curved shape and the side chains are in the proper positions for the molecule to function as a specific antibody, or as a specific enzyme, or, for that matter, in any other way.

Almost any sort of rough treatment, even gentle heat, is sufficient to break these feeble hydrogen bonds. When this happens, the polypeptide chain loses its proper shape and falls into disarray. Since the pattern of side chains has been destroyed, the protein molecule is no longer able to fulfil its function. It is this which accounts for the ease with which most proteins can be denatured and for the permanence of the denaturation.

The third stage in solving the protein pattern is to determine, then, the precise folding of the polypeptide chain. In English sentences this would be equivalent, once the exact order of the exact words is known, to establishing the exact context of the particular statement. Thus, the significance of the statement

John only punched Jim in his eye

is one thing if we are talking of two young prizefighters in the ring, quite another if we are talking of two elderly professors at a faculty meeting.

After Fischer had established the nature of the polyglycine backbone, chemists continued to tackle the problem of the protein pattern for a generation without making substantial progress.

It was not until 1935, as noted earlier, that the last amino acid was discovered. Even when all the amino acids were known, it remained impossible to solve even the first stage of the problem. A protein molecule could easily be broken down into a mixture of all the amino acids that made it up, but nothing any chemist could do in the 1930's would disentangle that mixture accurately. As late as 1944, in fact, there was still no way of telling with complete accuracy the number of each amino acid present in a particular protein molecule, and solutions to the second and third stages of the problem were entirely out of sight.

But in 1944, a sheet of absorbent paper came to the rescue.

The Pattern—Small Version

In that year, two English biochemists, A. J. P. Martin and R. L. M. Synge, devised a technique in which a mixture of amino acids, obtained by breaking down a particular protein molecule, was placed on porous filter paper and allowed to dry there. The edge of the paper was then dipped into an organic liquid that slowly moved up through the fibers by capillary action. (Dip a corner of a blotter into a glass of water, and you will see capillary action for yourself.)

As the liquid passed the dried spot of mixed amino acids, these amino acids found themselves pulled along.

Each amino acid was pulled at a different speed, and before long, each had been separated from all the others. Methods were easily devised thereafter for identifying each amino acid as it took up a characteristic position on the sheet of paper, and for estimating the quantity of each.

This technique, *paper chromatography* (kroh-muh-tog'ruh-fee), for the first time made possible accurate identification of all the amino acids in a particular protein. The first stage of the problem was solved and, through the late 1940's and thereafter, many exact identifications of the amino acids in one protein or another were performed.*

This was only the first stage. The attack on the second stage was launched at once, however. As soon as the Martin-Synge technique had been developed, another British biochemist, Frederick Sanger, tackled the problem of amino-acid order.

His method of attack was to break down the protein molecule only part way. Instead of reducing it to individual amino acids, he stopped at short lengths of peptides, each containing no more than two or three amino acids. He separated these small peptides by paper chromatography, isolated each one and worked on it separately. Carefully, he worked out the exact order of the amino acids in each of these small peptide chains (a delicate, but not impossible job), and then slowly deduced the manner in which all the amino acids must have been fitted together in long chains so that, when the latter were broken down, just those small peptides would be produced that he had actually detected, and no others. By 1953, he had worked out the complete amino acid order of a protein molecule called *insulin* (in'syoo-lin).**

The American biochemist Vincent du Vigneaud used Sanger's methods to work out the exact structure of two other protein molecules, called *oxytocin* (ok'sih-toh'sin) and *vasopressin* (vas'oh-pres'in). They turned out to be

* For developing the technique, Martin and Synge were awarded the 1952 Nobel Prize in Chemistry.
** For this feat, Sanger was awarded the 1958 Nobel Prize in Chemistry.

rather simple molecules, and he was therefore able to go a step beyond Sanger: he put together amino acids in the order he had deduced from his experiments. In that way, he was able to produce synthetic molecules that possessed all the properties, and performed all the functions, of the natural proteins. This was the strongest proof of the correctness of the theories of protein structure that had been developed from Fischer onward.*

And so the second stage of the problem had been solved as well, and we can now pause to examine the results. We can begin with vasopressin, one of the two proteins synthesized by du Vigneaud.

Vasopressin belongs to that class of compounds called *hormones*. It is produced by a specific organ (the posterior lobe of the pituitary gland, a small scrap of tissue at the base of the brain), and then discharged into the bloodstream. In small quantities, it—like all hormones—profoundly affects body chemistry. It increases blood pressure, for instance, but it also regulates kidney action, preventing the loss of too much water. When vasopressin is manufactured by the body in insufficient quantity, a disease called *diabetes insipidus* (dy-uh-bee'teez in-sih'pih-dus) results: a person suffering from this disease excretes a large volume of urine and suffers from constant thirst.

Du Vigneaud discovered that vasopressin obtained from oxen consists of eight different amino acids. These are, in alphabetical order: (1) arginine; (2) asparagine; (3) cystine; (4) glutamine; (5) glycine; (6) phenylalanine; (7) proline; and (8) tyrosine.

When he worked out their actual order, he discovered that the cystine molecule had its two amino-acid portions at different places in the polypeptide chain, so that part of the chain was bent into a loop, which was held in place by the disulfide link. He also found that the glycine molecule was at the unlooped end, and that the carboxyl group of that amino acid had been changed into an amide group. (Glycine, so modified, is called *glycinamide*.)

---

* This was so dramatic that du Vigneaud was awarded the Nobel Prize in Chemistry in 1955, the very year of his discovery, while Sanger had to wait three more years for the reward for his more general labors.

A simplified formula of ox vasopressin, with only the side chains showing, is shown in Figure 31.

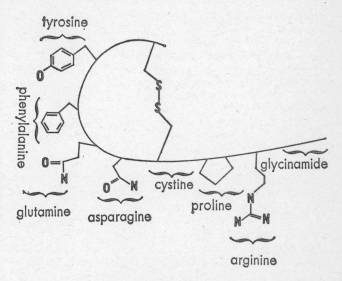

Figure 31. Ox *Vasopressin*

A second hormone produced by the posterior lobe of the pituitary gland is oxytocin, the other protein synthesized by du Vigneaud. It too consists of eight amino acids, of which six are identical with six in vasopressin. In place of the phenylalanine of vasopressin, however, oxytocin possesses an isoleucine, and in place of the arginine of vasopressin, it possesses a leucine. The simplified formula of oxytocin is shown in Figure 32.

Compare the two formulas in Figures 31 and 32 and you will see that the only difference is that the oxytocin molecule lacks a benzene ring and a three-nitrogen guanido combination, both of which are present in vasopressin.

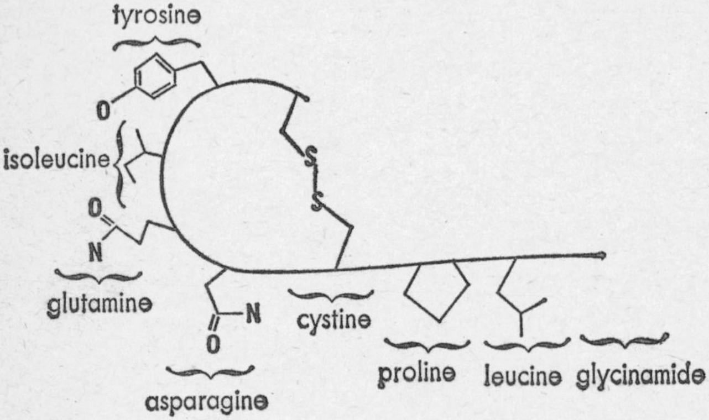

Figure 32. Ox Oxytocin

This may not seem like much of a difference, but it makes a world of change in terms of function. Oxytocin does not raise the blood pressure, as vasopressin does, nor does it have the latter's helpful effect on sufferers from diabetes insipidus. Instead, oxytocin brings about the contraction of the smooth muscles, particularly of the uterus, so that it is helpful in inducing childbirth. (Why the change of two side chains should produce such different sets of functions is as yet unknown.)

Changing two side chains out of eight is, however, a sizable alteration. It is possible to make a smaller change without loss of function. For instance, in the vasopressin obtained from pigs, seven of the eight molecules are identical with those in ox vasopressin, and arranged in the identical order, too. The only difference is that where ox vasopressin has an arginine, pig vasopressin has a lysine. To show this difference, it is necessary to write only the "tail portion" of the molecule: that is, the portion outside the cystine loop. It is shown in Figure 33.

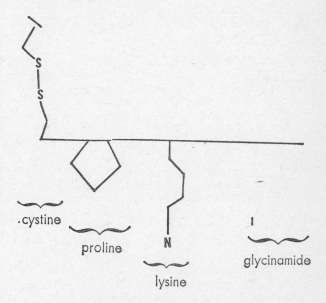

Figure 33. *Pig Vasopressin ("tail-portion")*

As you see, this difference is by no means as large as that between vasopressin and oxytocin. Pig vasopressin retains the benzene ring that is missing in oxytocin. Furthermore, it has not lost all three of the nitrogen atoms present in ox vasopressin but absent from oxytocin. In substituting a lysine for the arginine side chain of ox vasopressin, it still keeps one nitrogen atom in the side chain. The difference is small enough to leave the functioning of the molecule unaffected. Whether vasopressin is obtained from oxen or from swine, therefore, it will alleviate the sufferings of a patient with diabetes insipidus.

We can make an analogy between the three small

hormone molecules and the following three eight-word sentences:

1. *John Jones punched Mary Smith in the eye.*
2. *John Jones kissed Mary Smith on the eye.*
3. *John Jones kissed Mary Smith on the eyes.*

Two of the words in sentence 1 are changed in sentence 2, and the meaning is completely altered; the whole mood is different. John Jones switches from villain to hero, and Mary Smith's reaction would be different in each case. The first and second sentences thus represent the difference between oxytocin and vasopressin.

Sentence 3 contains one word—"eyes"—which is different from that in sentence 2. Nevertheless, the meanings of sentences 2 and 3 are essentially the same. That is the difference between ox vasopressin and pig vasopressin.

However, between sentence 2 and sentence 3 there is some difference, even if it is not enough to change the general meaning. Sentence 3 indicates at least two kisses and, therefore, implies a warmer relationship or a greater degree of privacy. In the same way, the chemical machinery of the pig pituitary produces something distinct from the product of the ox pituitary, and if the functions of the molecules are much the same in both species, the chemical machinery producing them must, nevertheless, be clearly different.

The Pattern—Large Version

A difference in structure cannot be ignored, even on a down-to-earth practical level, despite the fact that no difference in function is involved. I can show this best by turning to insulin, the first protein for which the second stage of the problem of structure was solved.

Insulin is a hormone formed by certain cells in the pancreas. Its presence controls the part of the chemical machinery of the body that handles the breakdown of sugar for energy. When it is present in insufficient quantity, the sugar breakdown slows down, and the serious disease *diabetes mellitus* (mel-ly'tus) results.

The molecule of insulin has a much more complex structure than that of either oxytocin or vasopressin. It contains a pair of polypeptide chains held together by two disulfide links. The two chains are called Chain A and Chain B: Chain A is made up of 26 amino acids, Chain B of 30. A portion of Chain A is forced into a loop by the disulfide linkage of a cystine molecule, as was the case with vasopressin and oxytocin. Within this loop, in addition to the cystine molecule itself, there are three other amino acids.

Insulin molecules obtained from the pancreas of a number of different species of animals have been studied and, if we exclude the disulfide loop, all are identical down to the last detail. Any change in the amino acids, or in their order (outside of that disulfide loop), seems to destroy insulin's function.

The three amino acids within the disulfide loop, though, can change from species to species without effecting any change in insulin's function. The changes are shown in Figure 34.

If these changes do not alter insulin's function, what is their importance? Granted, they may be of theoretical interest to the protein chemist, but do they have any *practical* importance? Oddly enough, they do.

Ordinarily, insulin does not give rise to much in the way of antibody formation. This is fortunate, because sufferers from diabetes mellitus require periodic injections of insulin, and it would be undesirable if their bodies reacted violently to the "foreign" but necessary protein. Occasionally, however, an individual may indeed develop antibodies against insulin obtained from cattle and thereafter be unable to tolerate the injections. In that case, it usually suffices to switch to insulin obtained from the pig pancreas. The difference in two amino acids out of fifty-odd isn't enough to change the insulin's function, but it is enough to generate the necessity for a different antibody. The antibody developed against ox insulin will not be required against pig insulin, and the patient can be treated once more, without the previous danger of adverse reaction.

Now it may seem to you that the actual pattern of amino acids is rather unimportant. One amino acid out of

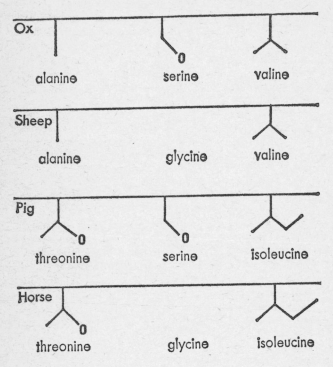

Figure 34. *The Various Insulins*

eight (12½ percent) can be changed in vasopressin without altering its function. So can three out of fifty or so in insulin (6 percent). Things seem a little flexible, anyway.

Well, a little—but not always even that.

Let's consider hemoglobin, the oxygen-carrying protein of the red blood cells, which I have mentioned once or twice before in this book. The normal hemoglobin molecules that occur in almost all human beings are called collectively, *hemoglobin A*.

There exist some individuals (fortunately, few) whose body machinery manufactures an abnormal hemoglobin. Two examples of such abnormal hemoglobins are *hemoglobin S* and *hemoglobin C*. The abnormal hemoglobins are not as efficient at picking up oxygen as is hemoglobin A. Furthermore, under certain circumstances, these abnormal hemoglobins will settle out as crystals within the red blood corpuscle, stretching, distorting, and damaging its membrane. Red blood corpuscles containing an abnormal hemoglobin therefore do not last as long as normal red blood corpuscles. Individuals who can manufacture hemoglobin A along with the C and S versions can still get along quite well; those who can manufacture only hemoglobin S or hemoglobin C are doomed to an early death.

Now the hemoglobin molecule is ten times as large as the insulin molecule. It contains a total of 574 amino acids, distributed among four polypeptide chains held together by disulfide links and electrical attractions. Two of these are identical "alpha chains," each with 141 amino acids; the other two are identical "beta chains," each with 146 amino acids.

In one of these chains (and in its pair) there is, in one certain position, a glutamic acid. If the glutamic acid is converted into a valine in each chain of the pair, the molecule becomes that of hemoglobin S rather than hemoglobin A; if the glutamic acid is converted into a lysine, the molecule becomes that of hemoglobin C. All the other five hundred and more amino acids remain (as far as is now known) identical in nature and in position. It would seem then that two amino acids out of 574 in one particular protein are all that stand between a healthy life and an early death.

There is no doubt, then, that the protein pattern is of vital importance—down to the last detail—and that "permission to vary" cannot be counted upon.

Since 1953, when the second stage of the problem was first solved, several other proteins have been worked out through this second stage—and more complex ones, too.

Vasopressin and oxytocin have only eight amino acids in their polypeptide chain; the longer of the insulin chains has only thirty. In 1960, however, the exact position of every amino acid was worked out for an enzyme called *ribonuclease* (ry'boh-nyoo'klee-ays), which possesses a chain made up of 124 amino acids, forced into a complicated series of loops by no less than four disulfide linkages stretching across from one spot in the chain to another.

There is no question that, given a large enough supply of a protein in its pure state, and given enough time and patience, the pattern of any protein molecule can now be cracked at least as far as the second stage.

What about the third stage? What about the bending of the polypeptide chain into a specific three-dimensional pattern, held in place by hydrogen bonds?

Even this has now been solved. Through the late 1950's, the English chemist John C. Kendrew, working with the Austrian-born chemist, Max Ferdinand Perutz, studied a protein called *myoglobin* (my'oh-gloh'bin), which is found in muscle. Like hemoglobin, it has the property of carrying oxygen; it is, however, only about one-fourth the size of hemoglobin. It is made up of one peptide chain and one iron-containing heme group, as against the four of each in hemoglobin. The single polypeptide chain of myoglobin, made up of about 150 amino acids, is not one of the hemoglobin chains broken loose, mind you. It is an entirely distinct structure.

Kendrew subjected crystals of myoglobin to X-ray diffraction studies (I shall have a few words to say about that in a later chapter), and was able to work out, little by little, the exact position of every portion. By 1959, he was able to prepare a three-dimensional model of the protein in which *every atom,* including the one iron atom, was in correct position.*

Given a sufficient supply of a pure crystalline protein of any sort, and given enough time and patience, it would seem that the same sort of model could be prepared. In short, all three stages of the problem of the protein pattern can now be regarded as solved, at least in principle.

* For this work, Kendrew and Perutz were awarded the 1962 Nobel Prize in Chemistry.

Much, of course, remains to be done; a very great deal remains. Nevertheless, protein chemists are riding a wave of optimism these days, and considering the tremendous progress that has been made in the less than twenty years since the introduction of paper chromatography, who can blame them?

The Pattern—in Potential

But is it possible that we are barking up the wrong tree with respect to the protein pattern? Granted that a change of an amino acid here or there makes a great difference, are there really enough changes possible to account for the endless variety of enzymes, antibodies, and so on? To be sure, if we go back to our usual analogy, there is an endless variety of possible English sentences. But English sentences are built up out of a vocabulary of some hundreds of thousands of words. What could the English language do if it had a supply of only twenty-two words?

On the other hand, English is limited in that words can be united only in certain combinations. You can say, "The grass being eaten by the ox is green." You cannot say, "The being by ox green grass eaten the is." At least, the latter is not a proper English sentence. In fact, almost any rearrangement of the words in a good English sentence is likely to generate nonsense.

In protein molecules, however, amino acids can be placed in any arrangement.

Let us see what this means by beginning with a simple eight-amino-acid protein, such as vasopressin or oxytocin. Suppose we number the eight amino acids: 1, 2, 3, 4, 5, 6, 7, and 8. Now, how many different arrangements are possible for these eight amino acids? Or, (which is the same thing), how many different numbers can be written using the eight digits from 1 to 8?

To begin with, you can start a number with any one of the eight digits. That's eight possibilities right there. Then as the second digit you can, for each of those eight, have any one of the seven remaining. This means a total of 8×7 or 56, counting only the first two digits. For each of these 56, there is a choice of any one of the remaining six as the third digit. The total for the first three digits is therefore $8 \times 7 \times 6$. If we continue, we can easily see that the total arrangements of eight digits (or of eight amino acids) is $8 \times 7 \times 6 \times 5 \times 4 \times 3 \times 2 \times 1$. The product of these numbers comes to 40,320.

This means that out of no more than eight amino acids in vasopressin, 40,320 protein molecules, each with somewhat different properties from all the others, can be built up.

The situation grows rapidly more startling as peptide chains grow longer. Suppose you had a peptide chain containing 30 amino acids, as in insulin. Of course, it would not consist of 30 different amino acids, and the presence of more than one of a particular kind of amino acid would cut down the possible number of arrangements. (Suppose, for instance, that the chain contained two glycines, one at position 4 and the other at position 14. If the 4 glycine was put at 14, and the 14 glycine at 4, it would still be the same molecule, for the two seemingly different arrangements would really be one.)

If we assume, then, that the 30-amino-acid peptide contains two each of 15 different amino acids (that is not quite the situation in insulin, but it is close enough), it turns out that the total number of different arrangements is about 8,000,000,000,000,000,000,000,000,000,000 or eight octillion.

Suppose we have a 140-amino-acid peptide similar to that in hemoglobin, and assume further that it is made up of seven each of 20 different amino acids. The total number of different combinations is something I won't bother to write down. In order to do so I would have to begin with the digits 135 and follow that with 165 zeroes. This is a far, far greater number than that of all the atoms in the known universe.

And so we answer our question. The number of different proteins that can be built up out of twenty-two amino acids is, for all practical purposes, unlimited. The amino acid side chains are quite enough to account for all the variety found in proteins; they are sufficient to form the basis for a phenomenon even as complex and subtle as life.

In fact, they are, if anything, more than enough. Out of 40,320 possible vasopressin combinations, the body chooses just one. Out of eight octillion possible combinations for one of the insulin polypeptides, the body chooses just one.

The question is no longer where the body finds the variety it needs, but how it controls the possible variety and keeps it within bounds.

It is to the search for an answer to that question that we must now turn.

CHAPTER *6*

Locating the Code

The Blueprint

If the cell is going to manufacture enzymes by forming one polypeptide chain—no other—out of virtually unlimited possibilities, there must be, somewhere in the cell, "instructions" as to how to go about doing this. It is inconceivable that a specific chain could be formed at random.

If, in ordinary life, a builder were told to construct the duplicate of a particular house, correct to the last nail, he simply could not do so just like that. He would either have to inspect both houses continuously, as he built, or he would have to have a detailed blueprint of the original in his possession.

In the cell, "construction by inspection" would involve the use of each different protein molecule as a model. A second protein molecule would have to be built upon the first, so to speak, amino acid by amino acid. But no biochemist has ever succeeded in causing an ordinary protein to produce a replica of itself, no matter what encouragement is given it in the way of raw materials, enzymes, special compounds, and so on.

The only bodies within the cell that do replicate are the chromosomes, and each individual begins life with chromosomes only. It must follow then that, within their own structure, the chromosomes possess the "blueprint" for the manufacture of protein.

This had been more or less the assumption ever since the chromosome theory of inheritance was adopted back in the early years of the twentieth century, and the assumption strengthened with the years. It was easy enough to talk of a "gene for blue eyes," but the gene itself didn't have blue eyes, and it didn't make blue eyes. It could only give the instructions for the production of a particular polypeptide chain that would then become the specific enzyme that would catalyze the production of a certain pigment that would give eyes a blue color. The end product might be a "physical characteristic," but the immediate job of the gene was to produce a particular protein.

It was not until the 1940's, however, that reasonably firm evidence was presented to back that assumption. Beginning in 1941, George W. Beadle and Edward L. Tatum began a series of experiments on bread mold. The wild strain of this mold could be made to grow on a medium containing sugar and inorganic salts. The salts included nitrogen compounds out of which the mold was able to manufacture for itself all the necessary amino acids. Not one ready-made amino acid had to be added to the medium.

Beadle and Tatum then subjected spores of the mold to X rays. As long ago as 1926, it had been shown by Hermann J. Muller that such radiation somehow altered genes and produced mutations, and Beadle and Tatum found that that was what happened in this case, too. Occasionally, an irradiated spore would refuse to grow on the ordinary medium, but it would grow if a certain amino acid—say, lysine—were added.

What had happened apparently was that the irradiated spore had lost the capacity to manufacture its own lysine out of the inorganic nitrogen compounds. Without lysine, it could not grow; if ready-made lysine were supplied to it, it could grow.

It seemed quite clear that some enzyme which ordinarily catalyzed one of the reactions that led to lysine was not being formed by the spore. It also seemed reasonable to suppose that this was because a particular gene had been damaged by the X rays. By carrying on a long series of ingenious experiments, Beadle and Tatum were able to make a strong case for the contention that every gene has as its task (and its only task) the formation of one specific enzyme. This is called the one-gene-one-enzyme theory.

There was considerable controversy over this when it was first announced, but most biochemists seem to accept it now. Actually, where an enzyme is made up of more than one polypeptide chain, a separate gene may be involved for each chain, so that perhaps it ought to be called the one-gene-one-polypeptide-chain theory.*

From this point of view the set of chromosomes with which each fertilized egg starts its life contains the information for a set of enzymes, roughly equal in number to the number of different genes. This "blueprint" in the chromosomes has, in the last few years, come to be called the *genetic code*. No phrase has swept so furiously through the scientific world since "atomic fission."

The Fall of Protein

But what do we mean when we say that chromosomes and genes carry the genetic code? How do they carry it? Of what sort of symbols is it composed?

The easiest assumption is that the code is written in the structure of the protein making up each gene. It seems safe enough to contend that only a protein molecule is complex enough to carry the blueprint for forming a protein molecule. Suppose each gene possessed somewhere in its structure the exact polypeptide chain of the enzyme whose synthesis it controlled. The gene would be a "reference enzyme," so to speak, carried on down from cell to cell and from organism to organism across the

---

* For his original work on X rays and mutation, Muller received the 1946 Nobel Prize for Medicine and Physiology. Beadle and Tatum, for their work on the irradiated mold, received a share of the 1958 Nobel Prize for Medicine and Physiology.

generations. By matching and rematching this reference enzyme, as many more enzymes of the same sort could be prepared as the cell needed.

All this seemed so natural that it almost went without saying—and yet there was evidence dating back half a century that seemed to point unmistakably to the fact that this view was incorrect.

In 1896, when scientists were just beginning to turn their attention to chromosomes, a German chemist named Albrecht Kossel was experimenting with salmon sperm which, like all other sperm cells, was merely a bag of compressed chromosomes.

He found, in the first place, that in content the salmon sperm was heavily nucleic acid; in fact the nucleic acid outweighed the protein two to one. And as if it were not enough that the protein was in such a minority, the protein molecules turned out to be rather unusual ones, which he named *protamine* (proh'tuh-meen). Protamine molecules are quite small for proteins and, curiously, are made up almost entirely of a single variety of amino acid. Some eighty to ninety percent of the amino acids contained in protamines are arginine.*

This is remarkable. When a macromolecule is so largely composed of a single unit, its capacity for carrying information is thereby drastically reduced. As an example, let us consider one peptide made up of ten different amino acids and another made up of ten amino acids of which eight are the same and only the remaining two different. The all-ten-different chain can be arranged in 3,628,800 ways; the eight-same-two-different chain can be arranged in only 90 different ways. Protamine molecules, then, have only 1/40,000 the potential for variety that a protein molecule of similar size but with a wider range of amino acids would have.

While it must be admitted that the capacity of variation in proteins is so vast that considerable reduction is permissible, it seems strange that this reduction should take place in the sperm cells, where the load of information to be carried must be at a maximum.

In ordinary salmon cells, the protein content of chro-

* Kossel was awarded the 1910 Nobel Prize in Physiology and Medicine for his research in this field.

mosomes is of a variety called *histone* (his'tone). This is a simple enough variety of protein, to be sure, but it is not nearly so simple as the protamines are. Why should a single body cell possess chromosomes with more complicated proteins than those possessed by the sperm cells out of which all the body cells would eventually develop? (And we need not suppose that the egg cell would make up the lack, since the father contributes no less to genetic inheritance than the mother, according to all evidence— and the father's portion of the inheritance is transmitted only by way of the sperm cells.)

Then, too, the enzymes in salmon (or in any other creature) are neither protamine nor histone, and therefore could not be copied directly from chromosom protein. And all of this holds for other species, too. Chromosomal protein, particularly in sperm cells, generally tends to be simpler than enzyme protein.

On the other hand, all tests on sperm cells since Kossel's time indicate that the nucleic acid of sperm is very similar to the nucleic acid of ordinary cells.

One way of solving this problem would be to suppose that, when an organism packs a set of its chromosomes into a sperm cell, all ballast that can be spared is thrown overboard. After all, the sperm cell has to whip its way to a waiting egg cell with all possible speed, and has good reason to travel light. This being so, what portion of the chromosomes would be conserved without change? Why, that portion which carries the genetic code. And which portion would be reduced and simplified as much as possible? Why, that portion which is not essential to the genetic code. From this point of view, it would seem that the genetic code is carried in the nucleic acid and not in the protein.

However, though this may seem rather obvious in the wisdom of hindsight, it was not at all obvious to chemists during the half-century following Kossel. Nucleic-acid molecules were then regarded as rather small and simple in structure. Even simplified proteins were considered more complex than nucleic acids, and to switch from protamine to nucleic acid would be to exchange bad for much worse.

Chemists therefore stuck with protein hoping that something would turn up to explain the too-simple protamine molecules and to show that they were, after all, complex enough.

But nothing came to save the protein-code theory. Instead, something was discovered that put it in its grave.

There are two strains of a certain type of pneumonia-causing bacterium. One strain grows a smooth, sugarlike pellicle over the bacterial cell; from the resulting appearance of the colony, it is called the "smooth strain," or S. The other produces no pellicle, is lacking in smoothness, and is called the "rough strain," or R.

In 1928, it was reported that a batch of dead S bacteria, killed by boiling, could be added to living R bacteria to bring about the production of living S bacteria:

$$\text{Dead S} + \text{Living R} \rightarrow \text{Living S}$$

It was inconceivable that dead S bacteria could come back to life, so one could only assume that the live R bacteria had been converted into live S bacteria, and that this had been brought about by something in the dead S bacteria.

The best guess seemed to be that the S bacteria possessed a gene that controlled an enzyme necessary for the formation of the pellicle. The R bacteria, on the other hand, lacked that gene, hence did not form that enzyme, hence did not possess a pellicle.

Dead S bacteria, however, still contained the gene. When the dead S bacteria were added to the living R bacteria, some of the R strain somehow picked up that gene and thus gained the ability to form the enzyme and then the pellicle. They became members of the S strain, in fact.

In 1931, it was shown that even intact dead bacteria were not needed to transform R into S. It was possible to make a bacterial extract that would do the trick. How? The extract contained the necessary gene.

Hope grew that the extract could be purified to the point where the gene itself could be isolated and studied. In 1944, this was accomplished and, like a thunderclap, the chemical nature of the gene was announced. Three biochemists working at the Rockefeller Institute, Oswald T. Avery, Colin M. MacLeod, and Maclyn McCarty, were able to show that the gene was nucleic acid—only nucleic acid. They were able to transform R to S by using a solution of the nucleic acid—without any protein at all!

Other examples of transformations from strain to strain were later reported among bacteria, and in every case the agent that did the transforming was nucleic acid. There was no blinking the fact: the genetic code could be carried by nucleic acid alone.

The Rise of Nucleic Acid

If any lingering doubt remained among biochemists, long used to thinking of protein as *the* characteristic stuff of life, it was snuffed out by experiments carried on virus molecules in the early 1950's.

After World War II, the electron microscope was developed to the point where it became both reliable and reasonably available to the average scientific budget. Since it made possible enlargements far beyond the capacity of any ordinary optical microscopes, virus molecules (too small to be seen by ordinary microscopes) became a field of study of wide interest.

It turned out that virus molecules generally consisted of a hollow shell of protein, within which was a nucleic-acid molecule. The latter was a single long structure, while the protein shell was constructed of a series of relatively small sections, all similar. Suddenly it began to seem doubtful, after all, that protein molecules were necessarily so much more complex than nucleic-acid molecules. Here was a clear-cut case in which the nucleic-acid molecule was larger than any of the protein molecules belonging to the system. (Of course, mere size may not signify complexity, as explained in an earlier chapter. I'll come back to this later in the book.)

In 1952, two biochemists, Alfred D. Hershey and M. Chase, performed a crucial experiment on *bacteriophage* (bak-tee'ree-oh-fayj), a variety of virus that infests bacterial cells. They enter a cell, multiply and become numerous, and finally kill the cell. The cell membrane bursts and, where one virus had entered, many emerge.

Hershey and Chase began by letting bacteria grow in a medium containing radioactive sulfur and phosphorus atoms. Since these atoms behave just like ordinary sulfur and phosphorus atoms, at least from the chemical standpoint, the bacteria incorporated the radioactive atoms into their own structures, just as they did ordinary atoms of each type. However, the radioactive atoms were continually breaking down and emitting tiny, energy-bearing particles that chemists could detect by the use of appropriate instruments. Thus, one could tell whether the particles were being emitted by the sulfur or the phosphorus. The bacteria grown in this radioactive medium were "tagged," so to speak.

The next step was to allow bacteriophage to infest these tagged bacteria. When this was done, the invading virus molecules made others like themselves out of the tagged interior of the bacterial cell; as a result, the newly formed molecules were themselves tagged. The tagging of the bacteriophage, however, followed a particular pattern. Protein molecules almost invariably contain sulfur atoms, but if they contain phosphorus atoms at all, these are very few in number. On the other hand, nucleic-acid molecules invariably contain phosphorus atoms, but never any sulfur atoms. As a result, a bacteriophage tagged with both phosphorus and sulfur atoms will carry the phosphorus in the interior or nucleic-acid portion, while the sulfur is found in the outer protein shell.

Finally came the crucial step. The tagged bacteriophage were allowed to infest normal, untagged bacteria. Now the presence of radioactive atoms would signify the presence of virus. Well, *only the radioactive phosphorus entered the bacteria*. The radioactive sulfur remained outside and could be washed off, or even simply shaken off.

The conclusion was unavoidable that it was only the nucleic-acid "innards" of the virus that entered the bacterium. The protein shell remained outside, discarded. And yet this virus nucleic acid, while within the bacterium, rapidly formed not only more nucleic-acid molecules like itself (although not like those normally found in the bacterium), but *also formed new protein shells*.

There could now be no getting away from the fact that,

in this case, at least, the genetic code is carried in the nucleic acid and not in the protein, and that the nucleic acid, without the protein, is capable of bringing about the formation of specific protein molecules. After all, the protein shell formed for the new virus molecules was just like the protein shell that had been discarded and left outside the bacterium, and unlike any of the proteins originally inside the bacterium.

Another blow was struck a few years later. In 1955, Heinz Fraenkel-Conrat developed gentle techniques for teasing the nucleic acid out of the protein shell of a tobacco-mosaic virus without damaging either the nucleic acid or the protein. Either part of the virus by itself seemed noninfective—that is, it could be smeared on tobacco leaves without causing the disease marked by the characteristic mottled discoloration. However, if the two parts were once again mixed together, some of the nucleic acid managed to work its way back into the protein shell, and the combination was once more infective. The following year, Fraenkel-Conrat was able to show that, while the protein did not seem able to infect the tobacco leaves, the nucleic acid—even by itself—did show a slight infectivity.

The meaning is clear. The protein shell acts, in the first place, as a "skeleton" designed to protect the nucleic acid portion of the virus. In the second place, the protein shell contains an enzyme which dissolves an opening in the bacterial-cell wall. (This dissolving enzyme was finally isolated in 1962.) The nucleic acid alone then enters the cell through that opening.

Without the protein shell, there is no enzyme to bore a hole for the nucleic acid, and the bare nucleic acid therefore seems to lack infectivity. Sometimes, though, the nucleic acid can manage, even by itself, to find a cranny through which it can wiggle, and then infectivity does show up, even in the absence of the protein.

The nucleic acid/protein combination would seem to be analogous to a man/automobile combination. A man and automobile together can travel from New York to Chicago without trouble. Either, separately, would seem incapable of doing so. The automobile, by itself, is obviously incapable; the man could, of course, if pressed by an overwhelming necessity, walk from New York to Chicago. There is no question that the man is the vital part of the man/automobile combination; by the same token, it is the nucleic acid that is the vital part of the nucleic acid/protein combination in viruses.

All experiments from 1944 on have pointed in the same direction. The nucleic acid, everywhere, in all species, in cells as well as viruses, was the carrier of the genetic code. The protein was never the carrier. Beginning with the late 1940's, therefore, chemists have eagerly turned to the nucleic-acid molecule.

And we must also turn to it, for we must now work out the structure of nucleic acid (as earlier we worked out the structure of protein) if we are to discover the nature of the genetic code.

CHAPTER 7

The Cinderella
Compound

Phosphorus

When nucleic acid came into its glory in 1944, it had been known for just about three-quarters of a century. During that time, however, it had been studied by only a handful of men. It was a Cinderella compound until, suddenly and quite without warning, the glass slipper was found to fit.

Still, those pioneers who had worked on it in the days of its obscurity had managed to deduce many of the basic facts of its structure. For instance, soon after its discovery it was found to contain phosphorus.

This was quite unusual. To be sure, some proteins were known to contain phosphorus, but only in small amounts. Casein, the chief protein of milk, is about 1 percent phosphorus. Lecithin, a fatty substance found in egg yolk, is about 3 percent phosphorus. Nucleic acid, however, is richer in that element than any other of the important substances of the body: it is 9 percent phosphorus.

It is time, therefore, to consider phosphorus in some detail. As noted in Chapter 3, the symbol of the element is P. In its chemical properties, phosphorus somewhat resembles nitrogen. Like nitrogen, phosphorus is able to combine with three different atoms. It can also upon occasion attach to itself a fourth atom (usually oxygen), by a special type of bond that is represented as a small arrow rather than a simple dash.*

As an example of this, Figure 35 shows the structural

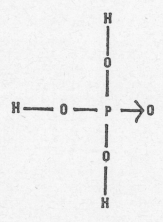

Figure 35. *Phosphoric Acid*

formula of *phosphoric acid* (fos-fawr'ik), an important industrial chemical. Its empirical formula, as you can see, is H_3PO_4. Notice also that whereas an oxygen atom can form two bonds of the ordinary "dash" type, it forms only one bond of the "arrow" type.

The bonds between the phosphorus atom and the oxygen atoms in phosphoric acid are strong. The hydrogen atoms can, however, be pried off the compound rather easily. If one is removed, an opening is made available for the attachment of what is left of the phosphoric acid

---

* Nitrogen can do this, too, but since its ability to do so plays no role in this book, I have not mentioned it.

to other atoms or atom groups. Two openings are made available if two of the hydrogen atoms are removed, and three openings if all the hydrogen atoms are removed. Phosphoric acid minus one or more of its hydrogens is a *phosphate group:* the phosphate is primary, secondary, or tertiary, depending on whether one, two, or three hydrogen atoms have been removed, making one, two, or three openings available for other combinations, as shown in Figure 36.

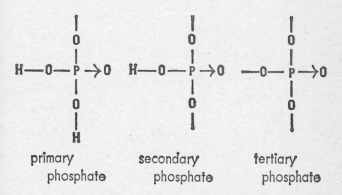

primary   secondary   tertiary
phosphate   phosphate   phosphate

Figure 36. *The Phosphate Groups*

In living tissue, the phosphorus atom always exists as part of either a primary phosphate or a secondary phosphate. It would simplify matters, then, to adopt a standard way of indicating these two groups, without worrying about their internal atomic arrangements. A convenient device is to let an encircled P represent, not phosphorus itself, but a phosphate group. We can differentiate between the primary and secondary phosphates by showing one or two bonds, respectively, on the symbol, as in Figure 37.

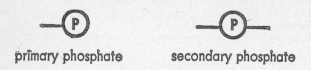

primary phosphate   secondary phosphate

Figure 37. *Phosphate Symbols*

As an example of the way in which a phosphate group may occur among the compounds we have already dealt with, consider the amino acid, serine. A phosphate group sometimes attaches itself to serine at the position of the oxygen in the side chain. In this way, *phosphoserine* is formed, as shown in Figure 38.

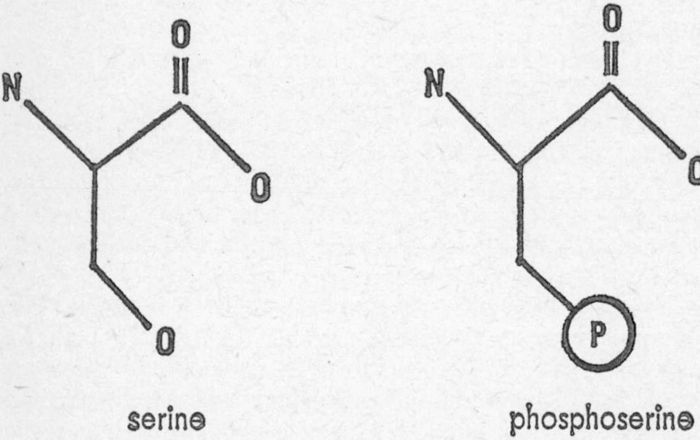

serine   phosphoserine

Figure 38. *Serine and Phosphate*

Phosphoserine occasionally occurs in proteins in place of serine and, when that happens, the result, naturally, is a phosphorus-containing protein or, as it is called, a *phosphoprotein*. Casein, which I mentioned at the beginning of the chapter, is an example.

So we can put phosphate groups down as one component of nucleic acid, and the component, moreover, that gives it its acid properties. It has, of course, other components as well.

The Two Varieties

Fairly early in the game, there were indications that nucleic acid contains sugar groups as part of its structure, but the nature of the sugar remained a complete mystery for decades.

The most common simple sugar in nature is glucose, the unit out of which starch and cellulose are built. The glucose molecule is a chain of six carbon atoms. To five of these a hydroxyl group is attached, while the sixth carbon atom makes up part of a carbonyl group. (It is this possession of a carbonyl group and numerous hydroxyl groups that is characteristic of the structure of a sugar molecule.)

Two other common sugars are *fructose* (frook'tose) and *galactose* (ga-lak'tose). Like glucose, each of them has six carbon atoms, one of which is part of a carbonyl group, while the others are attached to hydroxyl groups. Nevertheless, the relative orientation of the hydroxyl groups in space is different in each case. (Such a difference in orientation is quite enough to produce different compounds with different properties.)

Two simple sugars can combine with each other (just as amino acids can) with the elimination of water. Glucose and fructose combine to form a molecule of *sucrose* (soo'krose), the common "table sugar" we use to sweeten our coffee. Cane sugar, beet sugar and maple sugar are all sucrose. Glucose can also combine with galactose to form *lactose* (lak'tose), an almost tasteless sugar found only in milk. Finally, a number of glucose molecules can combine to form molecules of starch or cellulose.

There are numerous other sugars and sugar combinations. There are also some slightly modified sugar molecules: molecules to which nitrogen-, sulfur-, or phosphorus-containing groups have been added. Some compounds of this sort have never been found in nature, but have nevertheless been synthesized in the laboratory.

All these compounds—simple, combined, or modified; natural or synthetic—are lumped under the name of carbohydrates; as mentioned in Chapter 2, these make up one of the three major groups of organic materials in tissue.

But which carbohydrate is the carbohydrate in nucleic acid? The answer wasn't found until about 1910, when the Russian-born American biochemist Phoebus A. T. Levene first identified *ribose* (ry′bose) as a component of nucleic acid. Prior to that, ribose had not been known to occur in nature. It had been synthesized in 1901 by Emil Fischer (the man who worked out the peptide structure), but had been regarded as nothing more than a scientific curiosity of no practical importance. Even its name was simply made up by Fischer, with no particular significance intended. And yet it was destined to be recognized eventually as one of the two carbohydrates most important to life. (There are ugly ducklings which become swans in science as well as in ordinary life.)

Ribose differs from glucose, fructose, and galactose in that it has five carbons instead of six. This five-carbon chain tends to form a ring with an oxygen atom of one of the hydroxyl groups. The result is a four-carbon-one-oxygen ring that can be considered a furane ring without the double bonds. This is shown in Figure 39, where

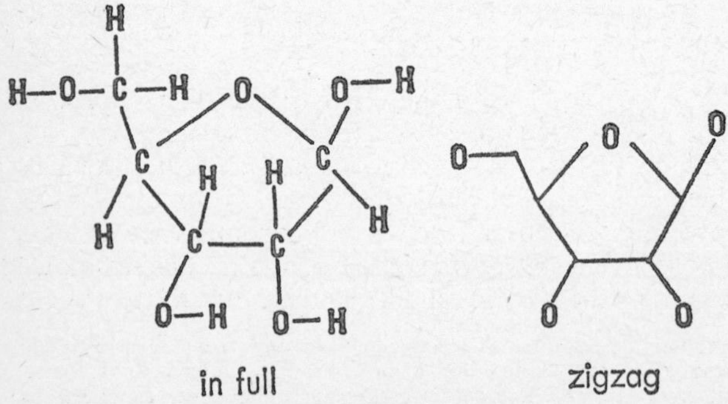

in full   zigzag

Figure 39. *Ribose*

ribose is presented both in full and in zigzag.

Later, Levene discovered that not all nucleic acid molecules contain ribose. Some samples contain a closely related sugar, which differs only in the absence of one of the oxygen atoms of ribose. Its name is, therefore, *deoxyribose* (dee-ok'see-ry'bose),* and its structure is shown in Figure 40. Deoxyribose, like ribose, had been

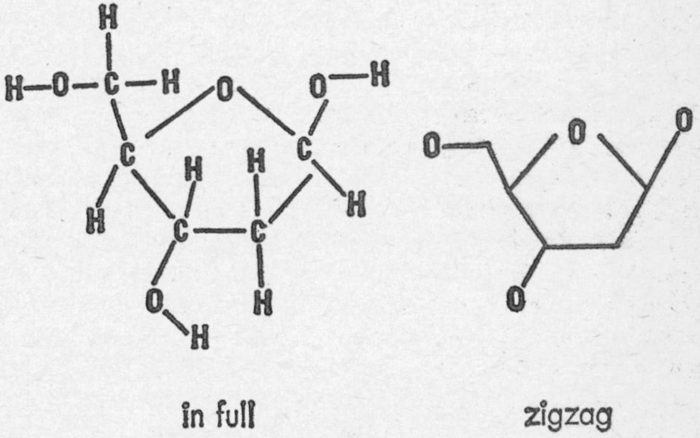

in full   zigzag

Figure 40. *Deoxyribose*

synthesized by Fischer years before it had been found to occur in nature.

It is on the basis of these two sugars that nucleic acid came to be divided into two types: *ribonucleic acid,* which contained ribose; and *deoxyribonucleic acid,* which contained deoxyribose. Since these names came to be used more and more often, and since biochemists are as allergic to polysyllables as anyone else, it soon became fashionable to refer to them by initials. Ribonucleic acid became *RNA*

---

* Until the middle 1950's, the prefix "deoxy" was universally written "desoxy" in the United States, and you may still find reference to "desoxyribose." By international agreement, however, American usage was switched to "deoxy" to bring it into line with usage in Great Britain and in other nations.

and deoxyribonucleic acid became *DNA*. Hardly anyone ever refers to them now except by initials.

No sugars other than ribose and deoxyribose have ever been found to occur in nucleic acids, and by the 1950's biochemists had settled down, more or less securely, in the belief that none ever would be, that ribose and deoxyribose are the only sugars in nucleic acid. Moreover, no nucleic acid has been found which contains both ribose and deoxyribose. It is either one or the other.

The two types of nucleic acid occur in different places within the cell. DNA occurs only within the nucleus and, indeed, only in the chromosomes. Some RNA is also to be found within the nucleus, but most of it is located outside, in the cytoplasm. All complete cells, as far as we know, contain both DNA and RNA.

As for viruses, the more elaborate ones contain, as cells do, both DNA and RNA. A good many, however, contain DNA only. The simpler ones, such as the tobacco-mosaic virus, contain RNA only.

Purines and Pyrimidines

In addition to phosphate groups and sugars, nucleic acids were found to contain atom combinations built around nitrogen-containing rings. This was discovered in the 1880's and afterward by Kossel (the man who was later to work with protamines). All the nitrogen-containing compounds that were isolated proved to be built about one of two ring systems, the purine ring and the pyrimidine ring, both of which were presented earlier in the book in Figure 15. The nitrogen-containing compounds isolated from nucleic acid are therefore lumped together as *purines* and *pyrimidines*.*

Two purines and three pyrimidines have been isolated in major quantities from nucleic acids. The two purines are *adenine* (ad'uh-neen) and *guanine* (guah'neen); the three pyrimidines are *cytosine* (sy'toh-seen), *thymine*

* Emil Fischer, who was later to work out the peptide structure, did considerable work on the chemistry of purines. For this, and for his work on sugars, he was awarded the 1902 Nobel Prize in Chemistry.

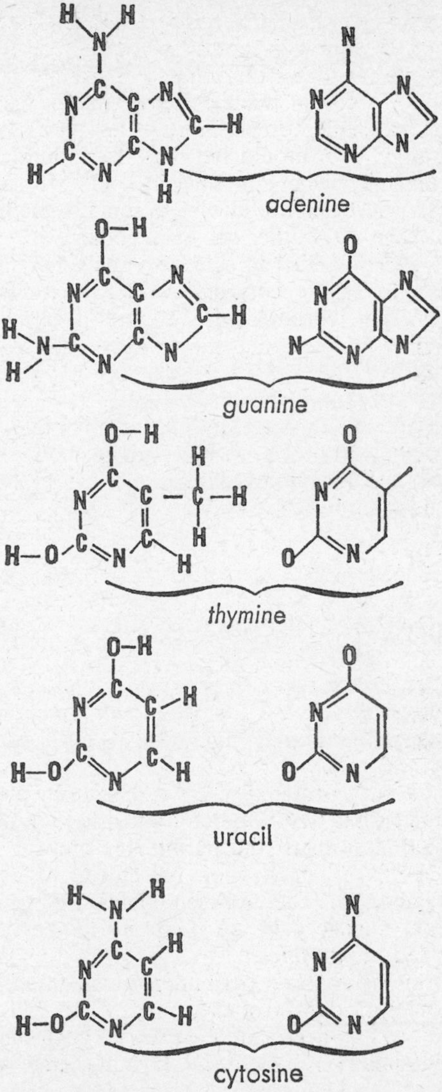

Figure 41. *The Purines and Pyrimidines*

(thy'meen), and *uracil* (yoo'ruh-sil). All five are presented in Figure 41, both in full and in zigzag.

Of these five, adenine, guanine, and cytosine occur in both DNA and RNA. Thymine occurs only in DNA, however, while uracil occurs only in RNA. Thymine and uracil are not very different; in fact, the only difference is that thymine possesses a methyl group which uracil lacks. In the zigzag formula, the thymine molecule shows a little dash where uracil does not, which really places the difference in proper perspective. As far as the genetic code is concerned (to get ahead of the story for a moment), the thymine in DNA is equivalent to the uracil in RNA.

One more word about the formulas. In certain organic compounds, it is possible for a hydrogen atom to move about rather freely, being attached to one atom at one moment and to another at the next. This takes place when double bonds are present, and the switch in the hydrogen atom involves a switch in double bonds as well.

In uracil, for instance, the hydrogen atoms of the hy-

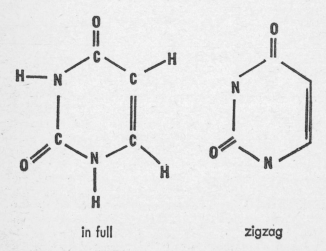

in full   zigzag

Figure 42. *Tautomeric Form of Uracil*

droxyl groups can easily switch to the nearby nitrogen atoms in the ring. In fact, they are rather more apt to be on the nitrogen atoms than in the hydroxyl groups. This phenomenon of the shifting of a hydrogen atom is called *tautomerism* (taw-tom'ur-iz-um). The tautomeric form of uracil is shown in Figure 42. If this is compared with the formula for uracil in Figure 41, you will see that, in the zigzag formulas at least, the only change is in the position of the double bonds.

(Actually, the phenomenon of tautomerism need not concern us any further. The only reason for mentioning it at all is that on occasion it is necessary to write the formula of a compound like uracil in one tautomeric form or another. Unless this is explained now, you may observe an unexpected difference in double-bond distribution from formula to formula and be disturbed by it.)

A couple of minor pyrimidines, with modifications of the cytosine structure, have been located in a very few samples of nucleic acid. Since these are all regarded as equivalent to cytosine, as far as the genetic code is concerned, you need not be troubled with them. The two purines and three pyrimidines listed in this section are all we will need.

Putting the Parts Together

Now we have the whole list. The phosphate group, ribose, deoxyribose, the two purines, and the three pyrimidines are the components of the nucleic acids. These make up eight "words," in comparison with the twenty-two "words" that make up the proteins.

This seems puzzling. Compounds that contain the genetic code should, it would seem, be at least as complex as proteins. Actually, things are even simpler than this. Of the eight "words," DNA lacks ribose and uracil, while RNA lacks deoxyribose and thymine. Each of the two varieties of nucleic acid, therefore, is built of only six "words."

But how are these "words" put together? Levene, the first to identify ribose and deoxyribose in nucleic acids, also tackled this problem. He broke down nucleic acid

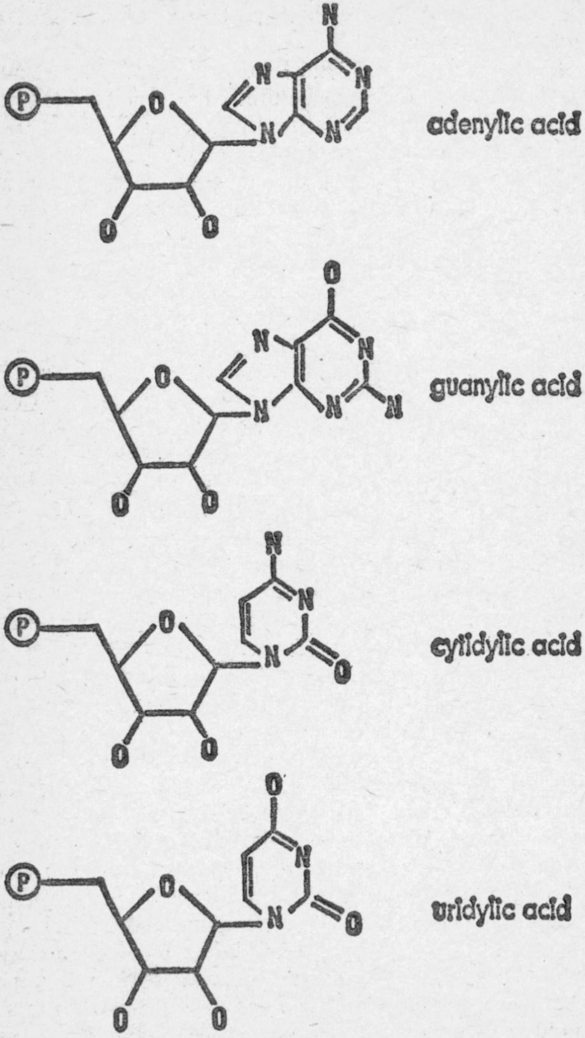

adenylic acid

guanylic acid

cytidylic acid

uridylic acid

Figure 43. The Ribose Nucleotides

into larger fragments, which contained several of the basic units. Working with these larger fragments, he deduced their structure.

In the early 1950's, the British chemist Sir Alexander R. Todd synthesized structures that followed Levene's suggested formulas and found that they did indeed possess the properties of the material obtained from nucleic acid. This was final proof of the suggestions that had, as a matter of fact, been fairly readily accepted by biochemists.*

What Levene had maintained was that, in the nucleic acid molecule, each ribose (or deoxyribose) portion has a phosphate group attached to one side and a purine or pyrimidine to the other. This combination of groups is called a *nucleotide* (nyoo'klee-oh-tide).

In RNA, all the nucleotides contain a ribose group, of course, and in addition either an adenine or a guanine or a cytosine or a uracil. There are thus four different nucleotides possible: *adenylic acid* (ad'uh-nih'lik, *guanylic acid* (gwah-nih'lik), *cytidylic acid* (sy'tih-dih'lik), and *uridylic acid* (yoo'rih-dih'lik). Again it is the presence of the phosphate group that gives each of these its acid properties, and adds the word "acid" to its name. And, of course, you can tell from the name of the nucleotide the particular purine or pyrimidine that it contains.

Since these nucleotides are of crucial importance to the genetic code, I will present the formulas for each, but in zigzag fashion only, in Figure 43.

The nucleotides in DNA differ in that they have deoxyribose in place of ribose. We can therefore speak of *deoxyadenylic acid, deoxyguanylic acid,* and *deoxycytidylic acid*. There is no deoxyuridylic acid in DNA: since the place of uracil is taken by thymine, there is *deoxythymidylic acid,* as shown in Figure 44. It differs from uridylic acid, as you can see, because of the missing hydroxyl group on the sugar. Deoxyadenylic acid differs from adenylic acid in the same fashion, and the identical difference appears when deoxyguanylic acid and deoxycytidylic acid are compared with guanylic acid and cytidylic acid, respectively.

* Todd was awarded the 1957 Nobel Prize in Chemistry for his work in this field.

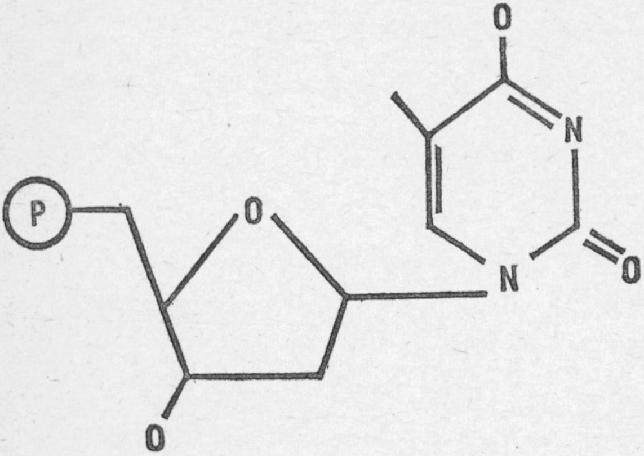

Figure 44. Deoxythymidylic Acid

Variations of this nucleotide arrangement are extremely important in the chemistry of the body. Nucleotides like those presented in Figure 43 exist but with the single phosphate group replaced by two, or even three, joined in tandem. They are key compounds in the storage and delivery of energy, the most familiar of them being a compound called *adenosine triphosphate* (uh-den'oh-seen try-fos'fate), usually abbreviated ATP. Its molecule is like that of adenylic acid, but with (as the name of the compound implies) three phosphate groups in place of the one possessed by adenylic acid.

There are also nucleotide-like compounds which work in cooperation with some enzymes, and are therefore called *coenzymes*. In these, the ribose portion is sometimes replaced by a glucose or some other carbohydrate, while in place of the purine or pyrimidine, there may be other types of nitrogen-containing rings.

However, we need concern ourselves here only with those nucleotides obtained from nucleic acids, and of these

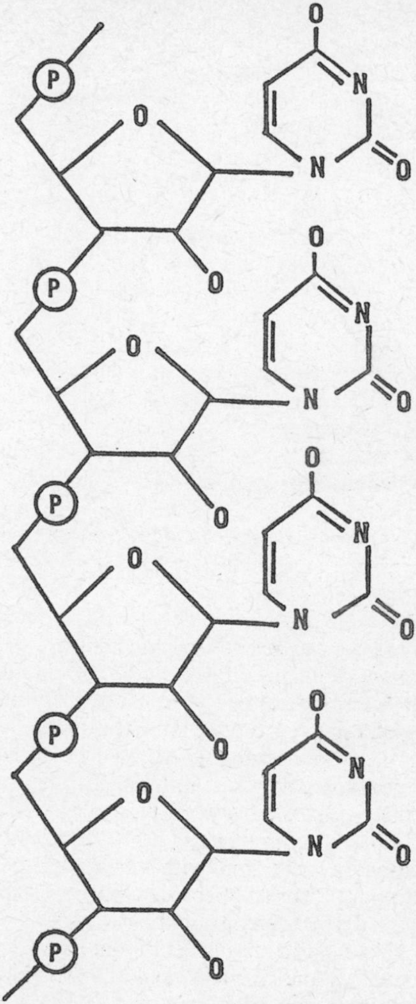

Figure 45. The Polynucleotide Chain

there are only four varieties in any one nucleic-acid molecule.

The next question we might ask is how the nucleotides are put together to form the nucleic acids themselves. This, too, was worked out by Levene and confirmed by Todd.

The secret lies in the phosphate group. In the individual nucleotides, it is usually a primary phosphate with one bond, although it can be a secondary phosphate with two bonds, and with the second bond attached to a second nucleotide. A whole series of nucleotides can be linked through secondary phosphates as is shown for a series of uridylic acids in Figure 45.

The joined nucleotides shown in Figure 45 make up a polynucleotide chain. Where the polynucleotide is made up of ribose nucleotides (as in Figure 45), each sugar group down the chain has a free hydroxyl group jutting out. (It is symbolized by the —O sticking out of each sugar ring.)

Where the polynucleotide is made up of deoxyribose nucleotides, this free hydroxyl group is not present. (Compare Figure 44 with Figure 43.) It follows then that RNA is made up of a polynucleotide chain with hydroxyl groups sticking out of the sugar portion, while DNA is made up of a polynucleotide chain without the hydroxyl groups.

The polynucleotide chain has a certain similarity to the polypeptide chain of proteins. The polypeptide chain is made up of a "polyglycine backbone" which runs the length of the chain and gives it unity; sticking out from it are the various side chains that lend the molecule its diversity. Similarly, the polynucleotide structure has a "sugar-phosphate backbone" which runs the length of the chain; sticking out from it are the various purines and pyrimidines. The schematic comparison is shown in Figure 46.

Only the side chains vary in the protein molecule, and only the purines and pyrimidines vary in the nucleic acid molecule.

Here arises what seems a serious paradox. There may be up to twenty-two different side chains along the polyglycine backbone (counting the absence of a side chain for a glycine itself as one of the items), but there are

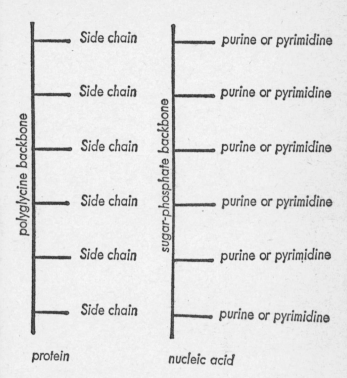

Figure 46. Protein and Nucleic Acid Compared

only four different purines or pyrimidines along the sugar-phosphate backbone.

How does the nucleic acid, with only four code-determining "words," supply the information necessary to build a molecule that may contain as many as twenty-two "words?"

We will get to this key question in time and find the answer to it, but only after we have looked quite a bit closer at the nucleic-acid molecule itself.

CHAPTER 8

From Chain to Helix

The Length of the Chain

Now that we know how the nucleotides are strung together within the nucleic-acid molecule, we can ask how many nucleotides make up a nucleic acid.

Until the 1940's that problem did not seriously engage the thinking of most biochemists. For one thing, it was taken for granted that the nucleic-acid molecule was fairly small. The mere fact that it was associated with protein made this seem reasonable: in any conjugated protein, the protein portion itself had to be (it seemed) the dominant member.

Consider hemoglobin, for instance. In addition to its 574 amino acids, it contains four heme groups. Each heme group is about five times as large as the average amino acid; even so, all four hemes put together make up only some 3 percent of the hemoglobin molecules. (The heme is called a *prosthetic group,* from a Greek word meaning "something added on.")

The heme is the "working portion" of the hemoglobin.

Each heme group contains at its center an iron atom to which oxygen molecules are loosely joined, so that hemoglobin acts as the oxygen carrier for the body. Nevertheless, it is the protein portion that actually determines the functioning of the heme group. There are various enzymes in the body: *catalase* (kut'uh-lays), *peroxidase* (pur-ok'sid-days), various *cytochromes* (sy'toh-kromez), each of which contains one or more heme groups; yet none of them can substitute for hemoglobin. All, in fact, have different functions, the differences being determined by the differences in the protein portion of the molecule.

There are other types of conjugated proteins, with other types of prosthetic groups. For example, there are *glycoproteins* (gly'koh-proh'tee-inz) with modified sugars as the prosthetic groups.

In every case, the prosthetic group seems to be a minor addition to the protein as a whole, with only a minor role to play. It therefore seemed natural to suppose that the nucleic acids, too, like other prosthetic groups, were comparatively small molecules, with some subsidiary function in the molecule as a whole.

To this "natural" assumption were added observations by Levene himself. He had isolated substances from the nucleoproteins which, on examination, proved to be nucleotide chains about four nucleotides long. They were, in other words, *tetranucleotides*. It seemed to Levene that these must represent the prosthetic group of the nucleoproteins. It seemed reasonable further to suppose that each tetranucleotide was made up of one each of the four different nucleotides.

Levene's conclusions were, unfortunately, based on observations that could not provide a true picture. His method of prying loose the nucleic acid from the protein involved the use of acids and alkalis. These pried the nucleic acid away, but they also broke up the nucleotide chain into small fragments. It was these fragments that Levene was studying.

Other biochemists eventually took to using milder methods of isolation, and they obtained different results. They isolated nucleic acids which consisted of chains much longer than four nucleotides. Slowly, the weight of evidence swung against the tetranucleotide theory of nucleic-

acid structure. Successively longer and longer chains were
obtained during the 1940's; by the 1950's, samples of
RNA were being obtained with molecules containing a
thousand nucleotides, and samples of DNA with molecules containing twenty thousand nucleotides.

These latest values are, if anything, rather on the long
side. It is quite possible that, in the present processes of
separation, several molecules of nucleic acid may be joining together loosely, and thus making the nucleotide chains
appear longer than they are in actual tissues.

At present, it is estimated that an individual gene may
consist of a nucleic-acid molecule made up of a chain of
anywhere from 200 up to 2,000 nucleotides.

The Diversity of the Chain

Even with the revelation that the nucleic acids may
be as large as proteins or even larger (a nucleic acid
made up of only 200 nucleotides is as large as a hemoglobin molecule), the tetranucleotide theory hung on for
a while in a modified version. It was granted that a
nucleic-acid molecule was more than four different nucleotides combined into one short chain, but now it was suggested that the molecule consisted of four different
nucleotides repeated over and over again to make one long
chain.

If the tetranucleotide theory, so modified, were correct,
then nucleic acids could not ever be expected to be the
carriers of the genetic code. Such a poly-tetranucleotide
would be simply a long "sentence" saying "and-and-and-
and—"

Just as the starch molecule is simply "glucose-glucose-
glucose-glucose—," so the nucleic acid would be simply "tetranucleotide-tetranucleotide-tetranucleotide—" The
fact that a tetranucleotide is about 7½ times as large as a
glucose molecule makes no difference. A sentence reading
"invincibility-invincibility-invincibility—" is not much more
informative than one reading "and-and-and—," even
though the word is longer and more impressive in the first
case than in the second.

Yet once the Avery, MacLeod, and McCarty experiments (see pp. 100–101) had been reported in 1944, biochemists began to realize, rather uneasily, that the tetranucleotide theory, no matter how it was modified, had to be wrong. Nucleic acid carried genetic information; the tetranucleotide model could not. Furthermore, as bacterial transformations were studied, it was found that nucleic acids existed in large varieties which could each bring about one particular transformation, but no others. This would not be so, if the tetranucleotide theory were correct.

Nucleic acids began to receive a closer and harder look.

Fortunately, in 1944, the same year in which Avery, MacLeod, and McCarty completely overturned all previous views about nucleic acids, Martin and Synge worked out the technique of paper chromatography. Although the technique was originally designed for amino acids, it was easily adapted for purines and pyrimidines.* The course seemed clear: break down nucleic acids, separate out the purines and pyrimidines, analyze this purine/pyrimidine mixture by paper chromatography, and then see if all four are present in equal quantities.

If all four are present in equal quantities, the tetranucleotide theory *may* be correct. According to the tetranucleotide theory, the purines and pyrimidines would be distributed 1-2-3-4-1-2-3-4-1-2-3-4—, so that there would be equal quantities of each. However, there might also be equal quantities of each distributed in a random order.

On the other hand, if analysis of the purine/pyrimidine mixture revealed the individual members to be present in unequal numbers, there would be no need to doubt —the tetranucleotide theory would be through.

And so it proved. One of the most assiduous investigators of this problem was Erwin Chargaff. By 1947, he was producing results that made it quite clear not only that purines and pyrimidines were present in unequal quantities within the nucleic acids, but also that the ratio of one nucleotide to another differed from one nucleic acid to another. The tetranucleotide theory was dead.

---

* As a matter of fact, paper chromatography could be, and was, adopted for almost any mixture of closely related substances, and in the few years since its development the technique has become an indispensable tool in every branch of biochemistry.

By the early 1950's, Chargaff was further able to show that the various nucleotides were, in fact, arranged in what seemed to be a random order. If this was so, then the number of different arrangements within a polynucleotide chain could be very great. Not as great, perhaps, as those within a polypeptide chain of similar size, for the polypeptide may have as many as 22 different units to shuffle, whereas the polynucleotide chain has only four.

Thus, for a polypeptide chain made up of 20 different amino acids, the total number of possible combinations is a little over 2,400,000,000,000,000,000 (nearly two and a half quintillion). On the other hand, for a polynucleotide chain made up of 20 nucleotides, or five of each of the four varieties, the total number of possible combinations is only a trifle over 1,100,000,000.

The polypeptide chain has the ability, in other words, to form over two billion times as many different combinations as a polynucleotide chain with a similar total number of units.

But who says that polynucleotide chains must have no more than the number of units possessed by a protein? Consider a particular polynucleotide chain containing twice as many nucleotides as there are amino acids in a particular polypeptide chain. The two can come up with roughly equal numbers of different arrangements. The limitation of having a maximum of only 4 different units, rather than 22, is compensated for by doubling the length of the more limited chain.

As it happens, the average nucleic-acid molecule contains perhaps five times (not merely two times) as many units as the average protein molecule. The disproportions in the way of forming different arrangements are therefore all in favor of the nucleic-acid molecule, after all.

By the early 1950's, there was no substantial doubt left that nucleic-acid molecules not only *could* carry the genetic code unaided, they *did* carry the genetic code unaided.

But why the nucleic acid rather than the protein?

It is always hazardous to ask "Why?" in science, but it is often interesting to do so just the same. Of course, we must remember that the answer to "Why?" is always a shaky thing, not to be compared with the strength and permanence of the answer to "What?"

In this case, my own speculation is that proteins are *too* complex and possess *too* many units. To store the blueprint of protein structure in a protein, and to expect it to keep its shape perfectly from cell division to cell division and from generation to generation of an organism, is perhaps too much. There are too many points where error can creep in.

Suppose, instead, that the information were stored in a polynucleotide chain. This has a sugar-phosphate backbone that consists of stacked rings of atoms, which is far sturdier than the relatively flimsy polyglycine backbone of protein molecules, a mere chain of atoms. Furthermore, the polynucleotide chain, with only four different units, offers the body a "choice" at each position of one of only four units, rather than one of twenty-two. The body is less likely to grow confused.

Enter the Helix

Even so, the question of how the genetic code is actually kept intact from cell to cell and from generation to generation is not so easily answered. Granted that a polynucleotide chain may be better fitted for the task than a polypeptide chain, the simple recognition of that fact still doesn't tell us just *how* the code is preserved.

The first step toward the answer arose out of the very investigations (into the numbers of purines and pyrimidines) that upset the tetranucleotide theory.

The inequalities among the purines and pyrimidines seemed at first blush to allow no hope for order. The number of adenine groups was usually higher than the number of guanine groups, for instance, but the amount by which it was higher varied with the species. In nucleic acids obtained from sea urchins, there were twice as many adenines as guanines. In human nucleic acid, there were only one and a half times as many. In some species, the situation was reversed and the guanine groups were more numerous than the adenines.

And yet as time went on, certain overall regularities were discovered, regularities that seemed to hold for all species and all creatures from man to viruses:

1. The total number of adenines seemed to be just about equal to the total number of thymines in DNA (or uracils in RNA), in the case of all the nucleic acids studied.

2. The total number of guanines seemed to be just about equal to the total number of cytosines, in the case of all the nucleic acids studied.

3. The total number of purines (adenine plus guanine) must therefore be equal to the total number of pyrimidines (thymine plus cytosine in DNA, or uracil plus cytosine in RNA).

These were interesting regularities and, as the event proved, important clues to nucleic-acid structure. Before proper use could be made of them, however, a crucial contribution was needed. It came in 1953, when the English physicist M. H. F. Wilkins studied nucleic acids by means of X-ray diffraction, and two colleagues, an Englishman, F. H. C. Crick, and an American, J. D. Watson, working at Cambridge University, used this work to advance an important theory of nucleic-acid structure. In the technique of X-ray diffraction (the technique later used by Kendrew to work out the exact three-dimensional structure of protein molecules) a beam of X rays is allowed to impinge on a substance. Most of the X rays pass through undisturbed, but some are diverted from their straight-line path.

If the atoms among which they pass are not arranged in orderly fashion, then the diversions are random. If the X rays are allowed to fall upon a photographic plate after passing through the substance, there is a central spot marking the position of the main beam. This has passed through undiverted and has survived to darken the photographic plate. About this central spot is a light fog caused by the effects of diverted X rays. This fog fades off continuously with increasing distance from the central spot, and is equally strong (or faint) at all angles from that spot.

If the atoms among which the X rays pass are arranged in orderly fashion, however, then the X rays are diverted in some directions more than in others. The orderly atoms lend each other a reinforcing hand, so to speak. This is most marked where the atoms are completely orderly, as in a crystal. A beam of X rays passing through a crystal will form a beautifully symmetrical pattern of dots radiating from the central spot. From the distances of these dots and the angles they make, it is possible to calculate the relative positions of the atoms within the crystal.

The same technique can be applied to macromolecules in which the units repeat themselves in some orderly fashion. Matters here are not quite as orderly as in a crystal, and yet neither are they completely disorderly. The X-ray diffraction pattern is fuzzier and harder to interpret, but it is not a featureless fog, and it is not impossible to interpret.

Watson and Crick, working backward from the X-ray diffraction data, came to the conclusion that the nucleic-acid molecule was arranged in the form of a helix. The helix is a figure with the shape of a circular staircase, often miscalled a "spiral" (so that one speaks of a "spiral staircase"). Actually, a spiral is a two-dimensional curve, something like a watch spring, while a helix is a three-dimensional curve, something like a bedspring.

This conclusion was not in itself a novelty. As pointed out earlier, the polypeptide chain can bend. Well, in 1951, the American chemists Linus B. Pauling and R. B. Corey were able to demonstrate that polypeptide chains in such proteins as collagen are arranged in helixes held together by hydrogen bonds.*

The Watson-Crick model of the nucleic-acid molecule, however, is rather different from the Pauling-Corey protein model. The Watson-Crick nucleic acid is made up of *two* polynucleotide chains forming an interlocking helix about the same central axis. The sugar-phosphate backbones make up the lines of the helix, while the purines and pyrimidines point inward toward the center, as shown in Figure 47.

It is this model which finally made sense of all the data

---

* For this, and for a great deal of excellent work done earlier on the bonds between atoms, Pauling was awarded the 1954 Nobel Prize in Chemistry.

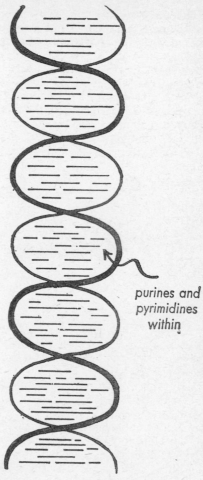

purines and
pyrimidines
within

sugar-phosphate
backbone 2

sugar-phosphate
backbone 1

Figure 47. The Nucleic Acid Double-Helix

that had been painstakingly collected on purine and pyrimidine ratios, and which was destined to make immediate sense of the problem of replication, as we shall see in the next chapter.*

* For their work in this field, Wilkins, Watson, and Crick shared the 1962 Nobel Prize for Medicine and Physiology.

CHAPTER 9

The Cooperating Strands

The Purine-Pyrimidine Match

The two helical strands of the nucleic-acid molecule are held together by hydrogen bonds between purines and pyrimidines at the point where the latter reach into the center of the helix.

Three arrangements are possible: a purine may be hydrogen-bonded to another purine; a pyrimidine may be hydrogen-bonded to another pyrimidine; or a purine may be hydrogen-bonded to a pyrimidine.

Since a purine is composed of two rings and a pyrimidine of one, a purine-purine combination will mean a long four-ring stretch from strand to strand; a pyrimidine-pyrimidine combination will mean a short two-ring stretch; and a purine-pyrimidine combination will mean an intermediate three-ring stretch.

If all three varieties of ring combinations—or even if any two of them—occurred along the double helix, then the two strands would be separated from each other by varying distances. The Watson-Crick model, as deduced

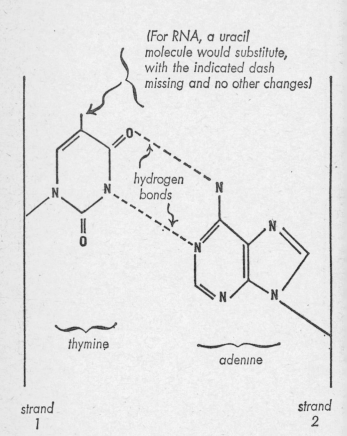

Figure 48. *The Adenine-Thymine Combination*

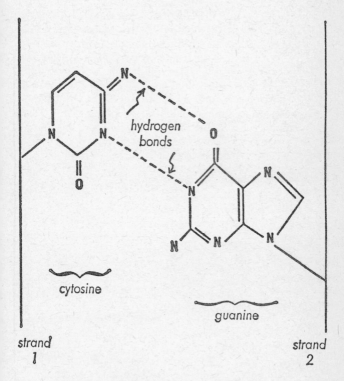

Figure 49. The Guanine-Cytosine Combination

from X-ray diffraction data, shows that to be impossible. The strands are separated by a constant distance all along the helix; the bonding must therefore be all purine-purine, all pyrimidine-pyrimidine, or all purine-pyrimidine.

But if the connection were invariably purine-purine, there would be no pyrimidines in the molecules; and if it were invariably pyrimidine-pyrimidine, there would be no purines in the molecule. Since no nucleic-acid molecule has been discovered in nature that does not have both purines and pyrimidines, purine-purine and pyrimidine-pyrimidine have to be eliminated as possible connections.

The only combination possible, then, is purine-pyrimidine. All along the helical strands, wherever a purine stretches inward from one backbone, a pyrimidine stretches out from a corresponding point on the other, and the two meet in the center via hydrogen bonds.

There are, of course, two different purines and two different pyrimidines, so there is still the question of which purine joins with which pyrimidine. That question is, however, easily answered. Since the number of adenines is found to be equal to the number of thymines (or uracils) in all nucleic-acid molecules analyzed, and the number of guanines is found to be equal to the number of cytosines, it is clear that an adenine must be hydrogen-bonded to a thymine (or uracil), and a guanine must be hydrogen-bonded to a cytosine, every time. Only in that way can strict equality be maintained.

The adenine-thymine combination is shown in Figure 48; the guanine-cytosine combination is shown in Figure 49.

It is interesting that in the adenine-thymine combination and in the guanine-cytosine combination, one of the two hydrogen bonds connects an N and an O. If thymine were to tie up with guanine, there would be an N—N hydrogen bond and an O—O hydrogen bond; if cytosine were to tie up with adenine, there would be two N—N hydrogen bonds. In neither of these "wrong" connections would there be an N—O hydrogen bond.

In short, so long as the distance between the sugar-phosphate strands remains constant, so long as both purines and pyrimidines are necessary parts of the molecule, so long as there are hydrogen bonds of the N—O type, we are certain to find the adenine-thymine (or uracil) and guanine-cytosine combinations—no others.

In this situation, the two strands within the nucleic-acid molecule are complementary. They are not identical; nevertheless, one matches the other by "opposites," so to speak. If we were to work out the exact order of the nucleotides in strand-1 of any nucleic-acid molecule, we would then be able at once to write the exact order of the nucleotides in strand-2 of that same nucleic-acid molecule. Where strand-1 had adenine, strand-2 would have thymine, and vice versa (or uracil in place of thymine for RNA). Wherever strand-1 had guanine, strand-2 would have cytosine, and vice versa.

For simplicity's sake, let us represent adenine by A, thymine by T, guanine by G, and cytosine by C. If the nucleotide succession in one DNA chain were ATTTGTC-CACAGATACGG, would you not know immediately that the nucleotide succession in the corresponding portion of the other chain was going to be TAAACAGGTGTCT-ATGCC?* You should. And in this respect, nature is at least as intelligent as we are.

Two for One

The Watson-Crick double-helix model of the nucleic acid proved immediately fruitful. Watson and Crick advanced the idea that, in cell division, the various nucleic-acid molecules making up the genes and chromosomes replicate themselves by a process in which each strand serves as a model for the other.

To make things simple, consider a DNA molecule made up of the usual double strand, but with only four nucleotides in each strand.

Strand A, let us say, consists of nucleotides containing an adenine, a cytosine, an adenine, and a guanine in that order: ACAG. Naturally, strand B must consist of nucleotides containing, in order, a thymine, a guanine, a thymine, and a cytosine: TGTC.

Now they separate. Strand A acts as one model. It

---

* This would hold for RNA if we substitute U (uracil) in place of T (thymine) at every point.

makes use of free nucleotides, which the cell can manufacture easily and which are therefore present at all times
in ample quantity and variety.

The first nucleotide in strand A contains an adenine,
which will automatically form a hydrogen bond with a
molecule of thymidylic acid. There is no "purposefulness"
behind this. Molecules will bump into the adenine through
the blind and aimless motion that is always agitating all
the molecules in the cell. With some of them, the adenine
may form hydrogen bonds. The strongest such bond,
however, will form when a thymine strikes in appropriate
fashion. The thymine will replace any molecule already
attached, and it will not be replaced by others. After a
period of time that is short by our human standards (a
thousandth of a second or less) but still long enough to
allow millions of collisions to take place, the thymine end
of the thymidylic acid is firmly in place.

In the same way, the second nucleotide in the strand,
which contains a cytosine, will form an attachment with a
guanylic acid. In short, the ACAG of the isolated strand
A will form a TGTC strand alongside itself. Meanwhile,
the isolated strand B will form an ACAG alongside its
own TGTC. In the place of the original double strand,
there will be two double strands, exactly alike, as shown
in Figure 50.

This Watson-Crick model of nucleic-acid structure and
replication is so clear and uncomplicated ("elegant" is
the adjective scientists would use) and explains so much
that other biochemists at once wanted to accept it. Scientists are human, after all, and a truly attractive theory
just begs to be believed.

And yet, however attractive a theory may be, it is always
better to have definite evidence supporting it.

Consider, then, that in the Watson-Crick model of
nucleic-acid replication, the individual polynucleotide
strands of DNA never break up. A pair of strands may
separate and attract free nucleotides out of which to build
a new strand, but the old strand remains intact throughout. When the cell dies, of course, all its polynucleotides
break up, but so long as life persists they do not.

Well, then, suppose an experiment is conducted that
will yield one result if the strands break up, another if

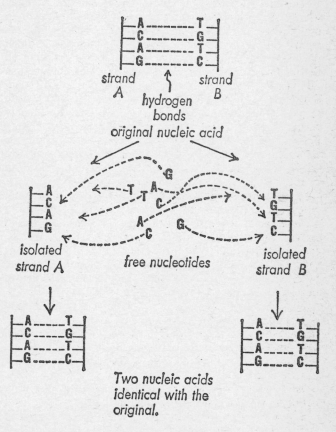

Two nucleic acids
identical with the
original.

Figure 50. Replication

they remain intact. This was done in 1958. Bacteria were grown in a medium containing great quantities of a heavy variety of nitrogen atom ("nitrogen-15," as compared with ordinary "nitrogen-14"), which can be easily distinguished from the ordinary variety by modern instruments. Gradually, the bacteria growing in this medium incorporated nitrogen-15 into the various compounds they were synthesizing and, in particular, into new polynucleotide strands. After the bacteria had been multiplying for a long time, virtually all the polynucleotide strands contained nitrogen-15. Each nucleic-acid molecule which contained two such strands could be called "15-15."

Now some of the bacteria with the "15-15" DNA were shifted to a medium containing the usual nitrogen-14 and were allowed to grow there for exactly two generations. What would you expect to happen?

If the polynucleotide strands broke up into small pieces, perhaps into individual nucleotides, and were then reconstituted, all the polynucleotide strands formed in those two generations would contain nitrogen-15. It would be diluted and thinned out by the influx of ordinary nitrogen-14 atoms, so that all the new strands would have less nitrogen-15, but every strand would still have *some* nitrogen-15. The nucleic acids would remain "15-15," and you could not distinguish one nucleic acid from any other.

But suppose the Watson-Crick picture were true, and the strands did not break up. At the first replication, each of the "15-15" nucleic acids would separate into two "15" strands. With each of these, as a model, new strands would be built up; they would contain only nitrogen-14, however, so that the new generation of nucleic acids—each formed of one old and one new strand—would all be "15-14."

With the second replication, the two strands of the new nucleic acids would again separate. This time, half the model strands would be "15" and half would be "14." All the new strands formed in this second replication, however, would be "14." Thus the third generation of nucleic acids would again fall into two categories, half of them being "15-14" and half of them "14-14."

After two generations, the nucleic acids were carefully tested, and two types, one with nitrogen-15 and one without, were indeed found. Similar results were obtained in experiments at Brookhaven National Laboratories, this time using growing plant cells and radioactive hydrogen. Eventually, some chromosomes proved to be radioactive, while some were not.

All this does not *prove* the Watson-Crick picture to be correct, but it certainly makes its correctness more likely. Had the results turned out the other way—had all the newly formed nucleic acids proven to be "15-15," for example—the Watson-Crick picture would have been completely and definitely shattered.

But it wasn't. As a matter of fact, all investigations in the years since Watson and Crick advanced their theory have tended to support it, and there are few biochemists now who do not accept it.

To be sure, a few viruses have been reported which seem to possess nucleic-acid molecules made up of but a single polynucleotide strand—and these viruses manage to replicate. Apparently, this is done by having the replication take place in two steps: the single strand produces its complement, and then the complement produces a replica of the original strand.

This is clearly less efficient than the usual double-strand method, for it means discarding half the strands that are formed. Though it works, the single-strand method seems to be restricted to a very few viruses. Most viruses and, as far as is known, all cellular creatures make use of double-strand replication.

The Watson-Crick model of replication implies that a polynucleotide strand may maintain itself intact throughout the lifetime of a particular organism. By chance, it may find itself in the egg cell or the sperm cell, and then it will continue onward into the new organism and go on for a new lifetime. In theory, it is even possible that, somewhere on the face of the earth, there are polynucleotide strands that have persisted through countless generations, perhaps even from the very first appearance of life.

This is, of course, unlikely. Most polynucleotide strands perish with the organism; only an insignificant minority find their way into the fertilized ovum and endure for another generation. It is likely that all the polynucleotide strands of this insignificant minority are secondary ones, formed in the course of the parent's lifetime. In that case, it may be that there actually are very few polynucleotide strands anywhere on the earth that are as much as a single century old.

Nevertheless, the possibility of a superpatriarch among the now-existing strands, straddling the eons since the earth was young, evokes a rather breathtaking picture of the unity and continuity of life.

Errors

Is replication always perfect? (Is *anything* always perfect?) Suppose strand A has a thymine in a particular position, and is all set to be joined by an adenine at that point. And suppose a guanine strikes the thymine in just the right orientation to form a hydrogen bond. It is possible that an adenine may not strike quickly enough to displace it, so that the line of nucleotides forms and joins together into a new strand, binding the misfit guanine in place.

In that case, you would not have a perfectly complementary pair of strands, A-B; instead, the new strand would be just slightly off, and we would have A-B'.

At the next replication, the two strands would separate. Strand A would form another strand exactly complementary to itself, for accidents are rare and not likely to happen twice in a row. Meanwhile, however, at this same replication, strand B' would form its own complement, A'. The misplaced guanine would attach a cytosine to itself, instead of the thymine that should be present in A.

This means that when nucleic acid A-B' replicates, it forms two different types of nucleic acids, A-B and B'-A'. Each type then perpetuates itself in future replications—barring, of course, the intervention of new accidents.

Nucleic acid B'-A' will not bring about the formation of the same enzyme as A-B. It is, after all, a different blueprint; the genetic code has been altered. The presence of a different enzyme will introduce a distortion into the chemical workings of the cell and we will have a mutation, the presence in the daughter cell of a characteristic not present in the parent cell. (It is just such a cellular mutation, some people think, that gives rise to a cell with a defective mechanism for regulating cell division. Such defective cells divide and divide, their numbers increasing indefinitely—and this is what we call cancer.)

If the new nucleic acid A'-B' finds its way into a sperm cell or an egg cell, from there into a fertilized ovum, all the cells of the new organism will possess it (barring further changes), so that the mutation will affect the new organism as a whole, and not just some of its cells.

A mutation may also be the result of the looping of strands in the process of replication. Perfect replication would require each strand to have all its nucleotide components available to the bombardment of free nucleotides, so that each component of the strand is able to pick up its proper complement.

Suppose, however, that a strand loops so that the components in the loop are placed out of action. A normal strand, with a CTAG section will have in its complement a section that goes GATC. If, however, the TA portion were looped out of the way and the C and G brought close together, a complement might be formed that was merely GC. Again, this abnormal strand would form a similarly abnormal complement at the next replication, producing a nucleic-acid molecule in which the looped TA portion is permanently lost.

The nucleotides of a resting strand of a nucleic-acid molecule may also be altered through reaction with particularly active substances in their neighborhood. Such changes will be made permanent through replication; once again, mutation.

Any factor in the environment that increases the probability of a mutation is a *mutagenic agent*. Heat seems to be mutagenic: as the temperature goes up, the mutation rate among bacteria or fruit flies (or other small organisms) rises. Perhaps this is true because even a small temperature rise weakens the feeble hold of the hydrogen bonds quite a bit. The difference in strength of the hydrogen bond between a nucleotide and its complement and the hydrogen bond between a nucleotide and its non-complement may decrease. It would then be that much easier to substitute a guanine where an adenine ought to go—that much easier to form a mutation.

Another mutagenic agent is radiant energy, which includes both the ultraviolet of sunlight and X rays, as well as the various radiations produced by radioactive substances. All of them produce *free radicals* within the cell. Such free radicals are fragments of molecules—usually, water molecules, since they far outnumber all other varieties in living tissue.

The free radicals are very reactive, combining with and altering almost any molecule with which they collide. If enough free radicals are formed, some are bound to collide with nucleic-acid molecules and alter them. The result is a mutation.

If the radiation dose is unusually heavy, the genetic code of vital cells may be damaged to the point where the cells can no longer perform their functions. This leads to "radiation sickness" and even death. It is this sort of danger that nuclear fallout presents mankind.

There are also chemicals which, by combining with nucleic-acid molecules and altering their structure, increase the rate of mutation. Of such chemical mutagens, the best known are mustard gas, of World War I notoriety, and related compounds called "nitrogen mustards."

Even under the best and mildest of circumstances, mutations will take place, for mutagenic agents cannot be completely eliminated. There is sunlight, constantly showering life forms with ultraviolet light. There are the radiations emanating from the radioactive substances present in tiny quantities in soil, sea, and air. There are the cosmic-ray particles bombarding us from outer space. And there are always the workings of sheer chance during the course of replication.

Accidents, in other words, will happen, and mutations will take place. For instance, there is a disease known as hemophilia, in which the blood will not clot, so that even a minor wound may cause someone who suffers from the disease to bleed to death. This is due to an "inborn error" in the chemical machinery of the body. The hemophiliac is born with an inability to manufacture some enzyme or enzymes that are essential at some point in the terribly complicated mechanism of clotting. Ordinarily, such inability to manufacture an enzyme (because of a defective nucleic-acid molecule in the chromosomes) is inherited. Nevertheless, it can also arise—through mutation—in a child of normal parents. Such a mutation will arise in an average of one out of thirty thousand births. (The mutation does not always show itself, by the way. For reasons I won't go into here, girls, but not boys, may possess the defective gene and still have blood that will clot normally.)

But mutations are not simply a matter of destructive error. Some changes may—through sheer chance—better fit an organism for its environment. It is upon this that the course of evolution through natural selection ultimately depends. Thus, a full century after Darwin constructed his theory of evolution on the basis of painstaking observations of organisms, scientists are substantiating that theory at the molecular level.

Man-made Strands

In nucleic-acid replication, the various free nucleotides must be joined together once they have taken their appropriate positions along the chain. This is apparently done in two steps. First, a second phosphate is added to the nucleotide—on the tail of the first phosphate, so to speak. The result is a "diphosphate." This second phosphate is then replaced by the neighboring nucleotide, and the two nucleotides are thus linked together by a secondary phosphate group. As this happens all along the line, a polynucleotide chain is built up.

Such a reaction must be catalyzed by an enzyme. In 1955, a Spanish-born American biochemist, Severo Ochoa, isolated just such an enzyme from bacteria. Adding this enzyme to a solution of the diphosphate variety of nucleotide resulted in a startling rise in viscosity. The solution grew thick and jellylike, a pretty good sign that long, thin molecules had been formed.

If one starts with a single type of compound, say *adenosine diphosphate* (the name given to adenylic acid that possesses a second phosphate group), then a long polynucleotide chain made up of a series of adenylic acids is formed. This is *polyadenylic acid,* or AAAAAAAA. . . . By beginning with uridine diphosphate, *polyuridylic acid,* or UUUUUUUU . . . can be synthesized. And so on. One can also begin with two, three, or four different diphosphates and end with polynucleotide chains containing two, three, or four components.

The building of the chains is slow at first; there is a sort of "lag period." After a while, when some of the chain has already been formed, that acts as the nucleus about which more can be formed, and the reaction is speeded up. The lag period can even be eliminated altogether, if some of the polynucleotide is added to begin with, as a "primer."

If one adds polyadenylic acid to a solution of adenosine diphosphate, additional polyadenylic acid is formed rapidly. If polyuridylic acid is added to a solution of adenosine diphosphate, however, the formation of polyadenylic acid is *not* hastened. The polyuridylic acid is the wrong primer.

Ochoa's work was with RNA. The next year, 1956, the American biochemist Arthur Kornberg did the equivalent with DNA. He isolated an enzyme that would form long polynucleotide chains out of individual deoxynucleotides, upon which three (not two) phosphate groups were to be found. Such nucleotides are "triphosphates" (adenosine triphosphate, or ATP, mentioned earlier, is an example).

Here, however, he did not form varieties of DNA made up of a single kind of nucleotide (at least not with this particular enzyme). Instead, the DNA chains were formed only when the four different kinds of deoxynucleotides were all present in solution. What's more, DNA was only formed when a sample of long-chain DNA was already present in the solution in addition to the triphosphates.*

Apparently, the formation of the two varieties of nucleic

* For this work, Ochoa and Kornberg shared the 1959 Nobel Prize in Medicine and Physiology.

acid proceeded in different fashion in the test tube. RNA was formed by the addition of one nucleotide to another, without the necessity of any guiding model. Primers were useful only as nuclei on which further nucleotides could be built up; and the chains built up were identical with, not complementary to the primer. DNA, however, seemed to be formed by replication even in the test tube.

This seems reasonable, since it is DNA, not RNA, that is the characteristic nucleic acid of the genes and chromosomes. It is DNA, not RNA, that is *the* replicating material in cells.

This is not to say that the RNA molecule cannot engage in replication, for it can. The proof of this is simply that a number of the simpler viruses contain only RNA, and no DNA at all. One example of this, as stated earlier, is the well-known tobacco-mosaic virus, the first virus to be crystallized. When a tobacco-mosaic virus infests a cell of the tobacco leaf, it multiplies within that cell, and new virus molecules by the hundreds are formed. Each new molecule contains an RNA molecule different from any of the RNA molecules in tobacco but identical with the RNA of the original invading virus. The new RNA molecules could only have been formed by replication.

Nevertheless, life based upon RNA replication is evidently not as successful as life based upon DNA replication. Only the simpler viruses are examples of the former; the more complicated viruses and all cellular life, without known exception, are based upon DNA replication.

And yet all cellular life retains RNA, in addition to DNA, and every species has characteristic varieties of RNA. How are specific RNA molecules built up and preserved through generations without replication?

The answer seems to be that RNA molecules can be built up by using DNA as a model. This had been assumed by many biochemists for years, but clear evidence for it was only presented in 1960. It was found then that DNA molecules were able to act as primers for the formation of RNA out of ribonucleotides, and even for the formation of an RNA molecule complementary to the DNA primer.

If DNA made up of a single type of nucleotide such as

polydeoxythymidylic acid (TTTTTTT . . .) is used as the primer, an RNA molecule is built up that also contains a single type of nucleotide. In this case it is polyadenylic acid (AAAAAAA . . .), since thymine and adenine are complementary.

If DNA made up of both deoxythymidylic acid and deoxyadenylic acid is used as a primer, then an RNA made up of the complementary adenylic acid and uridylic acid is formed. As nearly as can be told, the adenylic acids always form opposite the deoxythymidylic acids, while the uridylic acids form opposite the deoxyadenylic acids.

The formation of a complementary RNA takes place even when other nucleotides of a noncomplementary sort are available. In other words, if all four nucleotides are present in the solution, a DNA primer, made up only of deoxythymidylic acid, will still pick out only the adenylic acids and leave the others behind.

All this is not only evidence in favor of the manufacture of RNA on DNA models, it is also additional evidence in favor of the Watson-Crick model of replication.

The conclusion, then, is that DNA is the ultimate carrier of the genetic code in cellular life. If RNA carries the code, too, it is only because the information has been impressed upon it by DNA.

In that case, why is RNA needed at all? If it is merely a "copy-cat," what is it there for? Let's consider that next.

CHAPTER 10

The Messenger from
the Nucleus

The Uses of RNA

The importance of RNA was certainly not under-estimated in the days before the Watson-Crick model, even though it was known not to be the major component of chromosomes. If anything, it was overestimated because of the clear connection between RNA and protein synthesis.

The DNA concentration in the different cells of a particular organism seems to be a constant. Every cell, whether it is growing or not, whether it is constantly secreting material or not, has the same quantity of DNA. This is not surprising, since every cell has the same set of chromosomes, and it is there that the DNA is located. In fact, the sole exceptions are the egg cells and the sperm cells. They have only one of each pair of chromosomes, a half-set in other words, and to no one's surprise they contain only half the quantity of DNA contained in ordinary cells.

The RNA concentration in the different cells of a particular organism, however, varies over a wide range. Experiments dating back to the early 1940's have shown invariably that RNA concentration is higher where the rate of protein synthesis is higher. Growing cells are richer in RNA than are resting cells; a growing cell, after all, has to double its protein content between the time it is formed and the time when it is ready to divide again. When part of a tissue is growing and part is not, the RNA concentration is higher in the growing part.

Cells that form secretions rich in protein content— the cells of the pancreas and the liver, for instance—are also high in RNA. Furthermore, if enzymes which bring about the breakdown of RNA (but do not affect DNA) are added to the cell surroundings, so that the RNA molecules are in fact torn apart, protein production comes to a halt.

Put all this evidence together and there seems to be no doubt that RNA is deeply involved in protein synthesis. Such is the importance of protein synthesis to life that this evidence aroused a few suggestions in the very early 1950's to the effect that RNA was the more fundamental and vital variety of nucleic-acid molecule.

However, this pro-RNA position has not stood up. All the accumulated evidence makes it quite clear that it is DNA that is primary, and that RNA is a secondary product, so to speak, for which DNA is the model. This view is borne out by the fact that the RNA present in the chromosomes comes to less than 10 percent of the total quantity of nucleic acid, while there is a small structure within the nucleus (called the *nucleolus* [nyoo-klee'-oh-lus], from a Latin word meaning "little nucleus") that seems to be largely or entirely RNA. It seems reasonable to suppose that RNA is continuously being formed at the DNA sites in the chromosomes and then stored in the nucleolus.

Since protein synthesis takes place principally in the cytoplasm, RNA should be found there, and indeed it is. In fact, the major portion of cellular RNA is in the cytoplasm, although no DNA at all is present there. This means that RNA must be passed out of the nucleus and into the cytoplasm as it is formed. Studies with the electron microscope have actually photographed material bulging out from the nucleus and being pinched off in the cytoplasm, and these "blebs" (as they have been uneuphoniously named) contain RNA.

RNA, then, picks up the genetic code from the DNA of the chromosome and carries the message into the cytoplasm, where it supervises the formation of protein. Earlier, in describing the one-gene-one-enzyme theory, I said that it looked as if the only function of any particular gene was to form a particular enzyme. This is still true, but it must not be thought that this takes place in a single stage. Rather, the particular gene (DNA) forms a particular RNA, which in turn forms the particular enzyme. Perhaps it should now be called the one-DNA–one-RNA–one-polypeptide-chain theory.

It is easier to see the reason behind this double system of nucleic acids in the cell if we consider something similar in human technology. About a century and a half ago, the metric system was established, and for the first time science had at its disposal a truly logical system of measurements.

One of the fundamental units of the metric system was the "meter," originally defined as one ten-millionth of the distance from the equator to the North Pole, along the Paris meridian. However, that distance turned out not to be accurately known, so that the meter eventually came to be defined as the distance between two marks on a platinum-iridium bar, which was kept in a closely guarded air-conditioned vault in a Paris suburb.

That bar is called the "international prototype meter." Each country, as it adhered to the treaty establishing the metric system internationally, was given a copy of this standard, each copy being a "National Prototype Meter." In turn, each nation used its own national prototype to standardize the measuring rods which it manufactured for industrial, commercial, and technological purposes.

The national prototypes were well taken care of, for if any misfortune befell any ordinary measuring rod (or, worse, any of the calibrated machinery used in manufacturing such rods), the error could always be corrected by returning to the national prototype. And, if any accident befell the national prototype, even that error could be corrected by returning to the international prototype.*

The nucleic-acid situation is, it would seem, similar to this. The DNA is a "nuclear prototype," the equivalent of the international prototype in the metric system. It is therefore safely guarded in the nucleus, away from the rough-and-tumble world of the cytoplasm. The RNA molecules are the "cytoplasmic prototypes," which are of lesser importance, and the equivalent of the national prototype, or even of ordinary measuring rods. These can be risked in the hard task of protein synthesis.

We may even find a plausible reason now for the fact that DNA includes thymine, where RNA includes uracil. The actual difference between these two pyrimidines is as minor as one can imagine, resting on a single methyl group. Furthermore, this methyl group is located in such a position (see Figure 48) as not to interfere with the formation of hydrogen bonds with adenine. In DNA, adenine will link with thymine, while in RNA it will link with uracil; there seems no significant difference between these two bindings. In fact, when a molecule of DNA replicates itself, thymines will attach themselves at the adenine positions; when the same molecule of DNA produces an RNA, uracils will attach themselves there. There seems to be no difficulty in switching from one to the other.

It may be, then (and this is my own speculation), that the uracil serves merely as a "tag" for the RNA. The two nucleic acids, after all, suffer different fates. The DNA remains in the chromosomes at all times, while the RNA passes not merely out of the chromosomes where it is formed but out of the nucleus altogether. Whatever mechanism allows RNA to leave the nucleus and keeps DNA there must have some method of distinguishing between them, and the distinguishing factor must not be

*Presumably, something could happen to the International Prototype, so in 1960 it was internationally agreed to base the metric system on the length of the waves of light produced by a particular type of atom of the rare gas, krypton, when those atoms are heated. Now the measurements are nailed fast to an immutable (it is hoped) fact of nature.

one that interferes with nucleic acid functioning. Why not, then, as at least part of the distinction, the absence in RNA of a trivial methyl group which appears periodically in DNA?

The Site of Synthesis

If the cytoplasm is the place where RNA synthesizes protein, let's look at it for a while. The cytoplasm is by no means a smooth and homogeneous fluid; it is a complex system, containing thousands upon thousands of small bodies of various sizes, shapes, and functions.

The best known of these small bodies, or *particulates,* are called *mitochondria* (my'toh-kon'dree-uh), from Greek words meaning "granular filaments." The mitochondria are rod-shaped, with diameters of from 1 micron down to 0.5 micron, and lengths of up to 7 micra (a micron is equal to 1/25,000 of an inch). There are perhaps 2,000 mitochondria evenly distributed within the cytoplasm of an average cell.

In the late 1940's and early 1950's, methods were devised for separating the nuclei of cells from the cytoplasm and then separating the various types of particulates within the cytoplasm. As it became possible to study isolated mitochondria, it was discovered that they are the "powerhouses" of the cell. That is, virtually all the chemical reactions that produce energy by breaking down carbohydrate or lipid molecules go on within the mitochondria, which contain all the necessary enzymes and coenzymes for the purpose.

During the 1950's, more and more work was done with electron microscopes, which magnified the mitochondria sufficiently for scientists to discover that they are rather complex structures. Interest in them grew keen, indeed, and other particulates in the cell were lost in the shadow of the mitochondria, so to speak.

There were smaller particulates, for instance, called *microsomes* (my'kroh-sohmz), from Greek words meaning "little bodies," that were each about 1/10,000 the size of mitochondria. For a while, they were more or less ignored. It was even believed that they were merely fragments of mitochondria, broken off during the separation of the particulates.

There was, however, one fact which argued against this and created particular interest in the microsomes. This was the matter of chemical composition.

The mitochondria contain protein and certain phosphorus-containing fatty substances called "phospholipids." Together, these two types of substance make up almost all the material of the mitochondria. Very little nucleic acid is present in the mitochondria—only 0.5 percent of its substance is RNA.

Since the mitochondria are engaged in energy-producing reactions for which RNA is not needed, this is not very surprising. Still, the cytoplasm is rich in RNA; where is it, if not in the mitochondria?

RNA turned out to be located in the microsomes, which proved rich in nucleic acid. Since that was so, it wasn't very likely that they were merely fragments of mitochondria, the latter being poor in nucleic acid. Rather, the microsomes had to be independent particulates with an independent function. In view of their RNA content, might they not be the site of protein synthesis?

This hypothesis was borne out by experimental evidence. Cells supplied with radioactive amino acids incorporated the acids into polypeptide chains, so that the protein thereafter formed within the cell will turn out to be radioactive. If the cell is allowed to remain in contact with the radioactive amino acids for only a very short time and then immediately searched for radioactivity, only the proteins at the immediate site of protein formation should have been able to pick up the radioactivity. When this search was carried through, radioactivity was, in fact, found only in the microsomal fraction. The microsomes were therefore clearly the protein-factories of the cell.

The electron microscope now began to be focused on the microsomes. In 1953, the Romanian-born American biochemist George E. Palade found tiny particles densely distributed on the network of membranes associated with the microsomal fraction. By 1956, he had isolated these tiny particles (each about 1/10,000,000 the size of a mitochondrion, and perhaps not much larger than an individual gene), and found that they contained just about all of the RNA in the microsomal fraction. In fact, as much as 90 percent of the RNA in some cells is found in these numerous particulates, which are made up of RNA and protein, in about a 50-50 ratio. They came to be known as *ribosomes,* and through the late 1950's and early 1960's interest in them grew to the point where they actually put the mitochondria in the shade.

RNA at the Site

In the late 1950's, biochemists enthusiastically thought that in the ribosomes they had the important clue to the problem of protein synthesis. It was believed that each gene produced RNA by a Watson-Crick replication and that this RNA, after traveling into the cytoplasm, collected to form individual ribosomes.

This would mean that each different enzyme of the cell would be produced by some specific ribosome that had been formed by some specific gene. It was not believed that *every* ribosome would control a different enzyme—there were too many ribosomes for that. Rather, it was felt that a number of ribosomes would be responsible for this enzyme, a number for that enzyme, a number for the other enzyme, and so on.

This was made more plausible by the fact that a cell, under different circumstances, can produce enzymes at different rates of speed. Did this mean that, ordinarily, a cell would utilize only a portion of the ribosomes assigned to produce a particular enzyme, but that, in an emergency, more of the assigned ribosomes would be pressed into activity?

Unfortunately, difficulties arose. Sometimes an enzyme was formed at so high a rate of speed that one had to assume that a great many ribosomes were being placed into action—so very many, in fact, that it became unreasonable to believe that so large a fraction of the cell's ribosomes would be assigned to only one enzyme.

But what was the alternative? Suppose only a few ribosomes were actually assigned to produce that enzyme. Then, to account for the particularly rapid production of that enzyme, one would have to suppose that the individual ribosome was capable of multiplying its protein-forming capacity many times and reaching working efficiencies that were quite unbelievable.

Neither alternative sat well.

Another difficulty arose in connection with virus infestation of a cell. A virus-infected cell continues to manufacture protein at the same rate as an uninfected one, but the nature of the protein formed undergoes a change. The entry of a virus means an end to the formation of the cell's own protein and the beginning of the manufacture of virus protein. By the ribosome theory, this would seem to mean that when a virus enters a cell, it substitutes its own ribosomes for those of the cell. But considering how small a virus is, this seems impossible. A virus can hold only a very few ribosomes; how could these few outweigh the many thousands present in the cell?

Finally, there was the question of the *ribosomal-RNA* itself (the RNA molecules making up the ribosomes). It had a peculiar composition that weakened the whole idea.

DNA molecules, you see, vary markedly from species to species. Some species have DNA molecules that are rich in adenine but poor in guanine, the ratio being as high as 3 to 1; others have DNA molecules that are poor in adenine but rich in guanine, the ratio being as low as 1 to 3.

If the ribosomal-RNA is formed by the DNA of the chromosomes, it certainly ought to reflect these differences in base ratio—that is, if the Watson-Crick model of replication is correct. However, ribosomal-RNA does *not* reflect the varying ratio from species to species. In ribosomal-RNA, the four nucleotides are pretty evenly distributed, and so it was discovered in all the species of organisms tested.

Was the Watson-Crick model of replication wrong? Was the tetranucleotide theory right, after all? Biochemists couldn't bring themselves to believe that. They sought the explanation, and by 1960 they had found it. It seems that they had been following the wrong trail for three or four years.

Ribosomes are the site of protein manufacture all right, but ribosomal-RNA is not the means by which this is done. Ribosomal-RNA does not carry the genetic code; it merely serves as the structural backbone of the ribosomes. It is something like a "key-blank," which can be made to fit any lock, provided it is ground into the proper shape.

There must then be another variety of RNA: one which *is* formed by Watson-Crick replication from the gene; one which *does* carry the genetic code; one which travels from the gene to the ribosome with the gene's "message."

This second variety of RNA is called, appropriately enough, *messenger-RNA*. (It is sometimes called *template-RNA*, a "template" being a mold which serves as the guide for the production of some specific shape.)

Setting up the Key

Evidence for the existence of messenger-RNA reached the conclusive stage in 1960. Samples of RNA with a DNA-like distribution of purines and pyrimidines were isolated at the Pasteur Institute in Paris.

This DNA-like distribution was evidenced by the fact that the RNA could be bound by DNA strands from the bacteria which served as the source of the RNA, but not by DNA strands from bacteria of other species. The formation of a hydrogen-bonded union between a DNA strand and an RNA strand (a "hybrid nucleic acid") is possible only if the two strands are complementary. Presumably the RNA strand under investigation was complementary to a DNA strand from its own species of bacterium, because it had been formed from that DNA strand by replication.

Messenger-RNA must be formed from DNA at high rates of speed, for if the cell is briefly tagged with radioactive atoms, they show up at once in messenger-RNA. Then, after a brief time, the radioactive atoms appear, scattered elsewhere, in the cell. From this, it can be deduced that the messenger-RNA, once formed, is quickly broken down into individual nucleotides, which are then put to a variety of uses in the cell.

Messenger-RNA was first discovered in bacteria; indeed, a large number of contemporary discoveries in "molecular biology," as it is called, are made through experiments involving microorganisms. However, scientists in the field think that the findings are probably equally applicable to other creatures. In 1962, for instance, messenger-RNA was isolated for the first time from mammalian cells. Alfred E. Mirsky and Vincent G. Allfrey of the Rockefeller Institute obtained it from the thymus gland of the calf, and in much larger quantities than it can be obtained from bacteria.

Here, then, is the picture as it now stands:

1. The DNA of a particular gene manufactures a molecule of messenger-RNA by Watson-Crick replication. The messenger-RNA possesses a complement of the order of nucleotides in the DNA (except that there is uracil in all those places where thymine exists in DNA). Messenger-RNA, made up of perhaps as many as 1,500 nucleotides, thus carries the genetic code of the gene that made it.

2. The messenger-RNA molecule travels into the cytoplasm and attaches itself to an unoccupied ribosome. The "blank" of the ribosomal-RNA, now combined with messenger-RNA, is "keyed in" and becomes capable of manufacturing a specific protein. (In attaching itself, the messenger-RNA, it seems to me, must nevertheless leave its purines and pyrimidines free to form hydrogen bonds during the process of protein-manufacture—a process that will be described in the next chapter. It is my feeling, therefore, that messenger-RNA may attach itself to ribosomal-RNA by way of its own rear—that is, by forming hydrogen bonds with the hydroxyl group on the ribose units along the ribosomal-RNA chain. Perhaps this is why RNA possesses ribose instead of deoxyribose. The deoxyribose, and therefore DNA, lacks that free hydroxyl, as explained in Chapter 7, and perhaps RNA was "invented" just for the sake of that extra hydroxyl. Only so, one can suppose, is it able to function as a messenger.)

3. After a few protein molecules have been formed (or perhaps even after only one has been formed), the messenger-RNA breaks down, leaving the ribosome blank once again, ready to be "keyed in" for another protein,

perhaps the same one as before, perhaps a different one.

This whole process takes just two or three minutes, at most—which is amazing if you consider that in the course of the procedure, many hundreds of nucleotides must be precisely positioned to form messenger-RNA, and that many hundreds of amino acids must then be precisely positioned to form protein. On the other hand, you may be appalled at the thought of taking even a few minutes to form just a single protein molecule when so many are constantly needed at every instant of time. Of course, you can console yourself with the thought that there are millions of ribosomes per cell and that all of them, working together, can produce millions of protein molecules in those same few minutes.

The messenger-RNA picture removes the difficulties that plagued biochemists when ribosomes had been considered by themselves.

In the first place, it is no longer necessary to believe that each cell has special ribosomes for every different protein molecule it is capable of synthesizing. The ribosomes can be regarded merely as "blanks," which can be "hired out" temporarily to any DNA molecule. The purine/pyrimidine structure of the ribosomal-RNA therefore doesn't matter.

Furthermore, this means that the rate of enzyme synthesis can vary all over the lot, depending on the rate at which the various messenger-RNA molecules are formed. If one gene produces a great many messenger-RNA molecules, then these merely appropriate a correspondingly large number of ribosomal blanks and begin to manufacture enzyme at a correspondingly high rate. When the need is over, the messenger-RNA quickly breaks up, leaving the blanks blank again, ready for another task.

Then, too, the problem of virus infection and protein synthesis has grown less mysterious. The virus does not have to produce ribosomes of its own; it makes use of the bacterial ribosomes. (In 1960, experiments with radioactive atoms showed clearly that new ribosomes were *not* formed after viral infection.) What the virus does is somehow to bring about the suspension of the formation of messenger-RNA by the bacterial DNA. The messenger-

RNA already formed by the bacteria decompose with their usual rapidity, thus freeing the ribosomes, which can then be taken over by the messenger-RNA formed by the virus.

Protein synthesis is carried on at the same rate after infection as before, because all the ribosomes are still in use; they are merely coated now with virus messenger-RNA, and no longer with bacterial messenger-RNA.

Naturally, all this still leaves plenty of problems to keep biochemists busy. How does a particular DNA molecule "know" when to form a great deal of its particular messenger-RNA and when to form only a little? Apparently, the DNA must receive information, so to speak, concerning the state of its cell at every instant of time.

If a cell is short a component that it needs, the DNA molecules that govern the enzymes required for the formation of that component are somehow stimulated, and they produce greater quantities of their messenger-RNA. As a result, more enzyme is produced and more of the needed cell component is produced. If, on the other hand, a cell has a surplus of some component, the activity of the appropriate DNA molecules is suppressed.

This is a striking example of "feedback." The cell is quite obviously an extensive and complicated system of all sorts of feedback. The unraveling of the details whereby the DNA, RNA, enzymes, and products of enzyme-catalyzed reactions all interact upon each other will not be easy. Nevertheless, the problem is being attacked by biochemists with what can only be described as ravenous eagerness, and there is every hope that it will slowly yield before the onslaught.

CHAPTER *11*

Breaking the Code

The Triplets

Throughout the book so far I have deliberately avoided the key question in the whole business of protein synthesis: how does one go from nucleic acid to protein? Actually, by now we have narrowed it down to: how does one go from a specific messenger-RNA to a specific polypeptide chain?

At first glance, getting at the solution to this problem seems to be obscured by the same formidable obstacle referred to earlier. The nucleic-acid molecule is a "sentence" made up of four different "words," the nucleotides. The protein molecule is another "sentence" made up of twenty-two different "words," the amino acids. How can information carried by four different items suffice to explain what must be done by twenty-two different items?

This difficulty, which bothered many at first, is really no difficulty at all. That the thought even arises is only evidence that we are used to thinking of those codes in which a particular letter stands for some different letter,

as in the cryptogram puzzles one finds in some news-
papers. Thus, if in a particular cryptogram each letter
were represented by the next letter in the alphabet, the
word PROTEIN would be encoded as QSPUFJO.

And yet the most common codes we have by no means
work in this fashion. We have, for instance, exactly 26
letters in the English alphabet. These 26 letters are sufficient to encode the more than 450,000 words in Webster's
Third New International Dictionary (Unabridged). The
ten symbols used in constructing numbers (nine digits and
zero) are enough to encode an infinite set of numbers;
in fact, two symbols, 1 and 0, would be enough to serve
that same purpose, and do so in computers.

To make this possible, it is only necessary to agree to
make use of the coding symbols, the letters of the alphabet or the numerals, in groups.

There are only 26 letters to be sure, but there are
26×26 or 676 possible two-letter combinations, 26×26
$\times 26$ or 17,576 possible three-letter combinations, and
so on. In the same way, there are only 9 possible one-digit
numbers but 90 possible two-digit numbers, 900 possible
three-digit numbers, and so on.

In passing from nucleotides to amino acids, therefore,
we must abandon all thought of a one-to-one correspond-
ence and take the nucleotides in multiple units. There are
only 4 different nucleotides in messenger-RNA (or in the
DNA of the gene), but there are 4×4 or 16 different
dinucleotide combinations ("twins"), and $4 \times 4 \times 4$ or
64 different trinucleotide combinations ("triplets"). These
are all shown in Figure 51, where the four nucleotides are
represented by their initials: U for uridylic acid, C for
cytidylic acid, A for adenylic acid, and G for guanylic
acid.

This at once raises a new problem. There are too few
dinucleotides to account for the different amino acids,
but too many trinucleotides. We simply can't do with too
few, so that we have no choice but to go at least as far
as the triplets.

Now, it does not seem logical to suppose that we can
use some dinucleotides and some trinucleotides, having
just enough of both to make up twenty-two (or whatever

A   G   C   U

4 nucleotides

| AA | AC | GA | GC | CA | CC. | UA | UC |
| AG· | AU | GG | GU | CG | CU | UG | UU |

16 dinucleotides ("twins")

AAA	ACA	GAA	GCA	CAA	CCA	UAA	UCA
AAG	ACG	GAG	GCG	CAG	CCG	UAG	UCG
AAC	ACC	GAC	GCC	CAC	CCC	UAC	UCC
AAU	ACU	GAU	GCU	CAU	CCU	UAU	UCU
AGA	AUA	GGA	GUA	CGA	CUA	UGA	UUA
AGG	AUG	GGG	GUG	CGG	CUG	UGG	UUG
AGC	AUC	GGC	GUC	CGC	CUC	UGC	UUC
AGU	AUU	GGU	GUU	CGU	CUU	UGU	UUU

64 trinucleotides ("triplets")

Figure 51. *Nucleotide Combinations*

number of amino acids we settle upon). The trouble with that is that no one can figure out a way in which protein could "tell" when the combination AC, let us say, is a twin and when it is part of a triplet such as ACG.

If we did go on past the triplet stage and consider quadruplets, the situation would become even worse, for there are $4 \times 4 \times 4 \times 4$ or 256 different four-nucleotide combinations. It is therefore best to stick with the triplets —if we can.

There have been several efforts, however, to cut down the number of possible triplets so as to prevent what seemed to be the sad waste of allowing 64 triplets to stand for 22 amino acids. Suppose, for instance, that the nucleotide chain were read as a series of overlapping triplets, as in Figure 52. Such overlapping triplets (usually

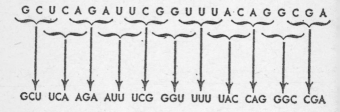

Figure 52. An Overlapping Code

following rather more complicated systems) could be designed in such a way as to cut down the number of possible trinucleotides to somewhere about twenty.

If you look at the overlapping code in Figure 52, however, you will see that every other nucleotide forms part of two triplets. The first U, for instance, is the last item of GCU, but also the first of UCA. Then there is an adenine which is both the last item in UCA and the first in AGA.

This sets up a serious limitation. In an overlapping code, such as that of Figure 52, the triplet GCU has to be followed by a triplet beginning with U, such as UCA. It cannot be followed, for instance, by AGA. Now if GCU stands for amino-acid-1, and if AGA stands for amino-acid-2, then this means that if the overlapping code is correct, amino-acid-2 can never follow amino-acid-1.

That sort of limitation is true of any overlapping code. An overlapping code, however arranged, must always limit the manner in which any one amino acid may be followed; there are always some "forbidden" sequences of amino acids in such a code.

Yet what we already know about the sequences of amino acids in proteins (see p. 81) is enough to tell us that there are no forbidden combinations. Any two-amino-acid combination is possible; any three-amino-acid combination is possible; and so on.

The result is that the code must be nonoverlapping— that is, in strict sequence, as shown in Figure 53. Here

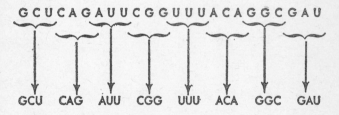

Figure 53. A Non-Overlapping Code

the triplets can come in any sequence, and so can the amino acids.

As usual, reasoning alone, however convincing, is all the better for having experimental support. In 1961, Crick himself, along with his coworkers, supplied such evidence. They began with an encoded nucleic-acid molecule which they were able to use to bring about a particular protein synthesis, and then added a nucleotide to it. The particular protein could no longer be synthesized. A second nucleotide was added; still no good. A third nucleotide was added, and the proper function of the molecule was restored. This could be interpreted, as shown in Figure 54, so as to strongly support the nonoverlapping triplet theory.

CUG:CUG:CUG:CUG:CUG:CUG:CUG:CUG:CUG CUG: original triplet sequence

add A CAU:GCU:GCU:GCU:GCU:GCU:GCU:GCU:GCU:GCU: triplet sequence changed

add another A CAU:AGC:UGC:UGC:UGC:UGC:UGC:UGC:UGC:UGC: triplet sequence still changed

add a third A ACA:UAG:CUG:CUG:CUG:CUG:CUG:CUG:CUG:CUG: original triplet sequence restored

Figure 54. Reconstituting the Triplet

That still leaves us, however, with 64 triplets for 22 amino acids. Two possible ways out still remain. Perhaps 32 different triplets are "blanks," and are therefore to be ignored in the general coding. Or it may be that two or even three different triplets all stand for the same amino acid. As we shall see, experimental evidence has definitely decided in favor of the second alternative.

A code in which two or more symbol combinations all stand for the same thing is said to be "degenerate." The genetic code, therefore, is in that class.

To summarize:

1. The genetic code consists of trinucleotide combinations, or triplets, running down the length of the polynucleotide chain, each triplet representing a particular amino acid.

2. The genetic code is nonoverlapping.

3. The genetic code is degenerate.

In addition, biochemists strongly suspect that:

4. The genetic code is universal; that is, that the same code holds for all organisms from the largest sequoia tree to the smallest virus.

The best evidence for the last statement is that a number of viruses are able to infect particular cells, each using its own messenger-RNA to produce proteins out of the cells' ribosomes, enzymes, and assorted chemical equipment. The cell can apparently "understand" the language of the various viruses. And in the laboratory, even when messenger-RNA from one species has been mixed with assorted cellular equipment from other species, the language has been "understood" and proteins were formed.

Making Use of Messenger-RNA

We can now picture messenger-RNA coating a ribosome and then directing the synthesis of a particular polypeptide chain through its sequence of triplets. But how is that done? It is all very well to say that a particular triplet "stands for" a particular amino acid; what makes particular amino acids actually line up in the order of the triplets?

The late 1950's brought the beginnings of the answer, chiefly through the labors of the American biochemist Mahlon B. Hoagland. In 1955, he discovered that, before being incorporated into the polypeptide chain, amino acids are joined to an adenylic acid. This combination is especially rich in energy and may be considered an "activated amino acid."

He then went on to discover the presence in cells of relatively small fragments of RNA, so small as to be freely soluble in the cell fluid. He called these fragments *soluble-RNA* as a result, but for reasons I shall shortly present, they are more often referred to as *transfer-RNA*.

It turns out that there are a number of varieties of transfer-RNA, and that each will attach itself to the adenylic acid portion of some activated amino acid. What's more, each will attach itself to one particular activated amino acid and to no other. What happens thereafter seems plain.

Let us suppose that a particular variety of transfer-RNA will attach itself to activated histidine and to that alone. The transfer-RNA will then transfer the activated histidine to the messenger-RNA (which is how transfer-RNA got its name). It will not, however, transfer to just any point on the messenger-RNA, but only to one specific point.

The transfer-RNA has a place of attachment, apparently, that consists of a particular triplet, and this triplet will only attach at that point on the messenger-RNA where the complementary triplet exists. In other words, if histidine's transfer-RNA has an AUG place of attachment, it will attach only to a UAC triplet on the messenger-RNA. In that way, the UAC triplet of messenger-RNA is attached, via transfer-RNA, to a histidine—and only a histidine. *Wherever* a UAC exists in the messenger-RNA, a histidine will be found, and that is how the triplet UAC can be said to "stand for" histidine in the genetic code.

This is borne out by an experiment carried out in 1962. A molecule of transfer-RNA was used which ordinarily attached itself to the amino acid cysteine. A technique was used whereby the cysteine was changed to the very similar amino acid alanine *after* it had been combined with the transfer-RNA. Despite this, the transfer-RNA with the alanine attached carried it into the spot where cysteine was ordinarily to be found. This showed that the attachment between transfer-RNA and messenger-RNA did not involve the amino acid, which had been changed, but only the purines and pyrimidines of the two varieties of nucleic acid, which had not been changed.

When all the transfer-RNA's are in place all along the polynucleotide chain of the messenger-RNA, the amino acids are dangling downward, in close proximity and in a particular order dictated by the triplet sequence of messenger-RNA (which it, in turn, obtained from the DNA of the gene). With the amino acids in close proximity and in proper order, it is easy for various enzymatic processes to bring about a reaction that combines them into one specific polynucleotide chain.

In 1961, Howard M. Dintzis of the Massachusetts Institute of Technology, working with amino acids tagged with radioactive atoms, devised experiments in which he was able to follow the appearance of radioactivity in proteins; he showed that the transfer-RNA's attached their amino acids down the chain of the messenger-RNA from one end to the other—in order, like strung beads.

This removed the possibility of confusion. Suppose you had the sequence AUUCGCUAG. In it the different possible triplets (if you could start anywhere) would be AUU, UUC, UCG, CGC, GCU, CUA, and UAG. Which of these seven triplets would be used by the transfer-RNA's, if they could grab on anywhere? Anarchy would result if one transfer-RNA should aim for UUC and another for UCG, so long as the two triplets overlap.

Instead, one transfer-RNA attaches to AUU. When that is done, another attaches to CGC, and when that is done, a third attaches to UAG. Any of the other four possible triplets are simply not used.

Dintzis was also able to show that all the amino acids in a molecule of hemoglobin can be set in place and bound together in a matter of about 90 seconds.

The whole scheme has been duplicated in the laboratory by using cell fragments rather than intact cells. In 1961, Jerard Hurwitz at the New York University Medical

Center set up a system containing DNA, nucleotides, and the appropriate enzymes and succeeded in manufacturing messenger-RNA in the test tube.

In that same year, G. David Novelli of Oak Ridge National Laboratories used not only DNA and nucleotides, but also ribosomes and amino acids. By doing this, he succeeded not only in manufacturing messenger-RNA, but also in having it coat the ribosomes and act as a model for the formation of a particular enzyme, called *beta-galactosidase* (bay'tuh-ga-lak'toh-sy'days).

The Triplet Dictionary

There still remains the question of the actual key of the code: which triplet stands for which amino acid?

The first breakthrough in this direction came in 1961 in what was perhaps the most important advance since the Watson-Crick model was proposed eight years earlier. The breakthrough was the result of an experiment by Marshall W. Nirenberg and J. Heinrich Matthaei at the National Institutes of Health.

They realized that in order to learn the key, it was necessary to start with the simplest possible situation— a nucleic acid made up of a chain of one single variety of nucleotide. Ochoa had already shown how such a chain could be built up with the help of the proper enzyme, so that polyuridylic acid, for instance, could be easily manufactured and used.

Nirenberg and Matthaei therefore added polyuridylic acid to a system that contained the various amino acids, enzymes, ribosomes, and all the other components necessary to synthesize proteins. Out of that mixture tumbled a protein that was as simple as the RNA they had in the beginning. Just as the nucleic acid was all uridylic acid, so the protein was all phenylalanine.

This was important. Polyuridylic acid could be represented as UUUUUUUUUUUU. . . . The only possible triplet that can exist in such a chain is, of course, UUU. The only amino acid used in building the polypeptide chain was phenylalanine, although all the different amino acids were present and available in the system. The conclusion that can be drawn from this is that the triplet UUU is equivalent to the amino acid phenylalanine.

The first step had been taken toward the decoding of the genetic code: "UUU means phenylalanine" was the first item in a "triplet dictionary."

The next step was seized upon at once; a number of research groups swung into action, following the lead that had been given them. Suppose a polynucleotide is built up enzymatically out of a solution of uridylic acid to which a little adenylic acid has been added. The chain will consist mostly of U, with an occasional A appearing at random. The chain may then be, for instance, UUUU UUUUUAUUUUUUUUUAUUUUUUAUUU. . . .

Such a chain would be made up of the following triplets: UUU, UUU, UUU, AUU, UUU, UUU, UAU, UUU, UUA, UUU. . . . The triplets are still for the most part UUU, but occasionally an AUU, UAU, or UUA will creep in. (These are the only three triplets that can be built from two U's and an A.)

Sure enough, the protein formed by such an "impure" polyuridylic acid turned out to be mainly phenylalanine, but with occasional "intrusions" of other amino acids. Three such "intruders" have been detected: leucine, isoleucine, and tyrosine. It seems clear that one of the three triplets AUU, UAU, or UUA stands for leucine, one for isoleucine, and one for tyrosine. Which is which, however, has not, at the moment of writing, been decided.

The best we can do is write UUA in parentheses (UUA), and permit that to signify the three different triplets that can be built from two U's and an A, without even trying to specify the order. In that case our dictionary could read: "(UUA) means leucine, isoleucine, or tyrosine."

If instead of adenylic acid, a little cytidylic acid or a little guanylic acid is added to the original solution of uridylic acid, polynucleotides are built up containing triplets that are (UUC) and (UUG). Again, the parentheses mean that we are not specifying the exact order of the three nucleotides.

In both these latter cases, leucine can still be detected in the still chiefly phenylalanine protein that is produced. This can only mean that (UUA), (UUG), and (UUC) can all be translated as leucine—an example of what we have called the "degeneracy" of the code.

If a small quantity of adenylic acid is once more added to the already "impure" uridylic-acid solution, so that the final polynucleotide contains a few more A's scattered among a profusion of U's, it is still very unlikely that any two A's will be close together. If this does happen by chance, however, the polynucleotide may contain such triplets as AAU, AUA, or UAA. These are the only three possibilities that can be made up of two A's and a U, and we can signify all three by (UAA).

As the quantity of adenylic acid added is increased, the (UUA) triplet will still increase in quantity, to be sure, but the (UAA) triplets will increase at an even faster rate. At first only the (UUA) triplets will be present in sufficient quantity to allow their amino acids to be detected. As the (UAA) triplets gain, however, new amino acids will begin to show up, which can then be attributed specifically to one or the other of the three triplets of (UAA). A similar situation holds if an increasing quantity of either cytidylic acid or guanylic acid is added. Amino acids matching the (UGG) triplets and the (UCC) triplets can be found.

What if both adenylic acid and guanylic acid were to be added in increasing quantities? At first only (UUA) and (UUG) are present in sufficient quantity for their amino acids to be detected. But then the various triplets represented by (UAG)—and there are no less than six of those—begin to mount in frequency, and new amino acids may appear and be attributed to them.

At the time I write, the correspondences that have been worked out between triplets and amino acids are as given in Figure 55:

The triplets represented in Figure 55 involve only 37 of the possible 64. The remaining 27 are those that do not contain U: for example, AGG, CCA, AAA, and so on.

Biochemists are confident that in a reasonably short time every triplet (defined in order as well as in content) will be lined up with a particular amino acid. There will then be a complete triplet dictionary, and the genetic code will be completely solved. For instance, by late 1962 there was already some evidence, obtained by Ochoa, that the triplet for tyrosine is AUU, in that order; the triplet for cysteine is GUU, in that order.

UUU	Phenylalanine
(UUG)	Cysteine Valine Leucine
(UUA)	Isoleucine Leucine Tyrosine
(UUC)	Leucine Serine
(UAA)	Asparagine Lysine
(UGG)	Glycine Tryptophan
(UCC)	Threonine Proline
(UCG)	Alanine Arginine Glutamine
(UAG)	Aspartic acid Glutamic acid Methionine
(UAC)	Asparagine Histidine Threonine

Figure 55. *The Triplet Dictionary*

CHAPTER 12

The Future

Subcellular Engineering

Attempting to look into a crystal ball is perhaps the riskiest of occupations. Unfortunately, it is also one of the most enticing. Given the chance to prophesy, only the strongest and most level-headed individuals can resist. I am not very strong in this respect and therefore I will try, with my fingers crossed, to peer into the future.

At the moment we stand at the beginning of what promises to be the most extraordinarily fruitful series of developments in the history of the life sciences. Problems that seemed insoluble twenty years ago have been solved; advances that seemed possible only in fantasy have become hard fact. And research is moving ahead at a rate and with a momentum that is higher than ever.

Biochemists have already used fragments of cells to manufacture specific proteins. There is no reason why this cannot be done for any protein. The ability to do so— an ability we possess *now*—is in essence a declaration of independence from life forms.

Consider the molecule of insulin. This is a substance that is necessary if we are to keep the disease of diabetes mellitus under control. Millions of diabetics depend upon it for normal life. At present, it is obtained from the pancreases of slaughtered cattle and swine. Enough animals are slaughtered for food purposes to supply the world with all the insulin it needs.

Suppose, though, that increasing population pressure forces future generations more and more into a vegetable diet. This would mean a steady decrease in the potential supply of insulin.

But what if we obtained a supply of insulin-producing cells from an ox pancreas, isolated the appropriate DNA and the ribosomes, and gathered the other necessary equipment? We could then, in effect, set up a "chemical plant," in which amino acids are fed in at one end and finished insulin emerges at the other, without any necessity of a live animal, or even of an intact pancreas, as intermediary.

To be sure, we would not have done away with the ox altogether. The original supply of DNA and ribosomes would still have come from a living pancreas. Yet, instead of having at our disposal only so much insulin as is in the cells at the time of slaughter, we could keep the subcellular equipment working for an indefinite period, and the amount of insulin obtained per pancreas would rise sharply. Our dependence on the ox would have narrowed considerably.

It might even be possible to arrange matters so that the DNA could replicate itself. Perhaps the day might come, then, when the pancreas would have to be rifled of its store only once. Thereafter, the system, properly cared for, would become self-perpetuating.

In fact, that day may be dawning, for in August of 1962, George W. Cochran of Utah State University announced that he had brought about the manufacture of a nucleic-acid molecule from the nucleotides, using various subcellular fragments and no intact cells at all. The nucleic acid produced was a good biological specimen, for it was the tobacco-mosaic virus nucleic acid, and Cochran had produced infective molecules.

Nor need insulin be the only protein so produced. There are many chemical reactions of industrial importance that are brought about by enzymatic means. Usually, this is done by taking advantage of the fermenting and synthesizing ability of bacteria, molds, and other microorganisms. Each microorganism, however, is busily engaged in a thousand reactions, which serve its own purposes, and which distract it from the one reaction in which we are interested.

If we could set up nucleic-acid/enzyme systems that would do the specific work required for that one reaction, we would have the equivalent of a super-specialized microorganism with no needs of its own, a single-minded molecular slave working for us indefatigably. A new specialized field of knowledge, "subcellular engineering," might arise, involving the preparation and control of such systems.

We might even learn to create new specialties. The nucleic acids we have can be delicately changed by treatment with heat, radiation, or chemicals, and the altered nucleic acids will then produce altered proteins. The vast majority of such proteins will undoubtedly turn out to be useless, but it is conceivable that once in a while an altered but useful protein (a "neo-protein") may be produced. Such a neo-protein may fulfill an old function more efficiently, or it may fulfill an entirely new one.

Just as at the present time there are specialists who carefully breed plants and animals in an eternal search for new and improved varieties, so there may someday be subcellular engineers whose chief concern is the everlasting search for new varieties of neo-proteins.

If we look far enough ahead, perhaps we can see a day when the production of neo-proteins need not be entirely hit-or-miss. If we learn enough about the structure of proteins, we may even reach the point where we can deduce what particular protein structure would be required to accomplish a particular purpose that is not accomplished by any of the proteins that already exist in nature. From our knowledge of the genetic code, we would then know exactly what nucleic acid is required to construct such a protein. Then, if we can but learn how to bring about the synthesis of even small quantities of such a nucleic acid, we are "in business" and can produce the new protein in large quantities.

In some ways, our situation is analogous now to what it was in, say, 1820. In that year, one might have predicted that chemists would learn how to construct organic compounds; that they would then proceed to construct thousands upon thousands of such compounds that were not found in nature; that they would even construct particular compounds which they knew in advance would have certain uses. In that year, one might have predicted that within a century and a half, synthetic dyes, synthetic fibers, synthetic plastics, synthetic pharmaceuticals, never yet found in nature, and far superior to any natural product for those uses to which they were to be put, would come into common use. Such predictions, however, would have sounded irrationally fantastic.

Now we can predict the same thing, but on the subtler, more intricate, and more wonder-working level of protein chemistry. Is this also irrationally fantastic?

The Ultimate Goal

The prospects for the future are not only a matter of new chemical industries. Knowledge begets knowledge, and the promise of current research in molecular biology is fabulous.

If a particular messenger-RNA is isolated in quantity, and if the enzyme it controls is identified, that messenger-RNA might then be used to identify the particular DNA molecule that formed it. It would attach itself to that portion of an isolated chromosome which would be its exact complement and to which it could then attach strongly via hydrogen bonds.

The way would then be open for precise "chromosome mapping." Naturally, this would not be easy. However, a beginning is being made. In 1962, Robert S. Edgar of the California Institute of Technology announced that he had located about half the genes present in a particular virus, working out the nature of the enzyme each produced. To be sure, he didn't use messenger-RNA for this purpose, but older techniques involving mutations. Also, the virus has only 100 genes altogether, whereas a man may have as many as 150,000. Still, it *is* a beginning. One might in the end identify by such means every DNA molecule in every chromosome.

Thereafter progress might take place in a number of different directions. For instance, the chromosomes of cells in different tissues might be mapped in order to help solve the frustrating problem of what makes one tissue different from another.

After all, even an organism as complex as a human being begins as a single fertilized cell with but one double set of genes. Fifty trillion or more human cells arise from that original and, though this sounds like an enormous proliferation, they can arise by means of no more than 47 successive cell divisions.

You can check this by considering that the original cell will become two cells after one division. The two cells will become four after a second division, and the four cells will become eight after a third. Carry this through 47 times, if you have the patience, and note the final number.

In each case the chromosomes replicate, so it is to be expected that all the cells of the human adult body have identical genes. Yet that would mean that all must have identical enzymes, too, and therefore identical cellular machinery and identical properties.

The fact of the matter, however, is that they don't. The cells of each organ and of each tissue of an organ have their own characteristic enzymatic makeup, their own abilities, their own properties. A nerve cell, a muscle cell, a bone cell, a kidney cell, a salivary gland cell are each 47 replications removed from the same fertilized ovum, and yet how different they are from each other.

The chemical basis of this differentiation of tissue is only now slowly beginning to be understood. Until very recently it was not known whether different tissues arise because, in the course of cell division, certain groups of cells actually lose particular sets of genes, or whether each contains complete sets of genes but neutralizes or inhibits the action of certain of them.

At least two lines of very recent experiments seem to support the latter of the two alternatives. At Oxford University, researchers have been killing the nuclei of frog eggs by ultraviolet light. Nuclei from frog embryos or even from tadpoles that have newly hatched have then been substituted. Thirty percent of the nuclei from the embryos sufficed to allow the division of the egg cell and produced normal adult frogs. Four percent of the nuclei from the intestinal cells of a newly hatched tadpole did the same. Apparently, then, even after considerable differentiation, a frog cell nucleus contained all the genes necessary to produce a complete frog.

The work of Ru-chih C. Huang and James Bonner at the California Institute of Technology fits in here. They studied the protein components of chromosomes and found that in some cases they could increase the rates at which messenger-RNA was produced by removing certain varieties of protein present in the chromosome. It is possible, then, that some proteins serve as "locks," inhibiting the action of certain nucleic-acid molecules. In that case, every cell, however specialized it might be, could still contain every gene, but each might possess its own pattern of locking protein which would block out certain genes in nerve cells, other genes in muscle cells, and so on.

If all this turns out to be so, then it is conceivable that we may also learn to unlock the genes. Is it possible, then, that we may someday be able to encourage the stump of an amputated arm to grow a new whole one by de-differentiating its cells and then allowing them to grow and differentiate once more? Can we obtain scraps of embryonic tissue or fertilized ova and *direct* them to the production of hearts only, or of kidneys, as these are needed for transplantation?

Nor need we stop merely with the possibility of physical repairs. We might be able to correct overall imperfections: counteracting hormonal imbalances, or canceling out altogether the possibility of cancer.

The precise points of deficiency in various inherited diseases and in the disorders of the cell's chemical machinery may be spotted along the chromosomes. This could lead the way to early diagnoses of conditions which ordinarily develop only later in life. It might even become possible to spot the existence of such a defect, where it is suppressed in a given individual by the presence of a normal DNA molecule on the paired chromosome. That defect, you see, prevented from reaching open expression in the individual himself, may turn up in the next generation in full expression.

One may speculate about some far future in which individuals will routinely undergo "genic analysis," as nowadays they are routinely vaccinated. This could lead eventually to the development of a rational basis for eugenics—that is, for a course of action designed to remove deleterious genes and encourage the dissemination of desirable ones.

Perhaps massive genic analysis of the population will eventually give us the information that will lead to working out the physical basis for mental disease. We might even work out the gene combinations for such things as high intelligence, artistic creativity, and for all the things that are the essence of humanity in its highest and most idealized form.

Will the day come, then, when we can reach the ultimate goal of directing our own evolution intelligently and purposefully toward the development of a better and more advanced form of human life?

INDEX

SIGNET SCIENCE LIBRARY Books
60¢ each unless otherwise noted

- [ ] **NEW HANDBOOK OF THE HEAVENS by Hubert J. Bernhard, Dorothy A. Bennett, and Hugh S. Rice.** A guide to the understanding and enjoyment of astronomy for beginners as well as the more advanced, with star charts and data, descriptions of the heavenly bodies, and astronomical facts and lore. (#Q3647—95¢)

- [ ] **SEEING THE EARTH FROM SPACE by Irving Adler.** A timely, up-to-date report on Russian and American satellites and what we are learning from them about our earth. Illustrated. (#P2050)

- [ ] **THE SUN AND ITS FAMILY by Irving Adler.** A popular book on astronomy which traces scientific discoveries about the solar system from earliest times to the present. Illustrated. (#P2037)

- [ ] **THE STARS by Irving Adler.** A clear introduction to the nature, motion, and structure of the stars. (#P2093)

- [ ] **MAGIC HOUSE OF NUMBERS by Irving Adler.** Mathematical curiosities, riddles, tricks, and games that teach the basic principles of arithmetic. (#P2117)

- [ ] **THINKING MACHINES by Irving Adler.** How today's amazing electronic brains use logic and algebra to solve a great variety of problems. (#P2065)

- [ ] **THE NEW MATHEMATICS by Irving Adler.** The first book to explain—in simple, uncomplicated language—the fundamental concepts of the revolutionary developments in modern mathematics. (#P2099)

- [ ] **MATHEMATICS IN EVERYDAY THINGS by William C. Vergara.** In fascinating question and answer form, and illustrated with diagrams, this book shows how the basic principles of mathematics are applied to hundreds of scientific problems. (#T2098—75¢)

- [ ] **THE NATURE OF THE UNIVERSE by Fred Hoyle.** A noted astronomer explains the latest facts and theories about the universe with clarity and liveliness. Illustrated. (#P2331)

- [ ] **SATELLITES, ROCKETS AND OUTER SPACE by Willy Ley.** A newly revised and up-dated report on the science of rocket development, including an evaluation of the flights of Titov and Glenn, and of the satellite, Telstar. (#P2218)
- [ ] **HOW LIFE BEGAN by Irving Adler.** A readable account of what science has discovered about the origin of life. Preface by Linus Pauling. Illustrated. (#P2135)
- [ ] **EARLY THEORIES OF THE UNIVERSE by James A. Coleman.** From the earliest recorded observation of the movements of celestial bodies through the major contributions of Brahe, Kepler, Newton and their successors, a noted scientist and writer provides the layman with a history of man's progress toward a scientific understanding of the universe. (#T3137—75¢)
- [ ] **THE CRUST OF THE EARTH edited by Samuel Rapport and Helen Wright.** Selections from the writings of the best geologists of today, telling the fascinating story of billions of years in the life of the earth. (#T3138—75¢)
- [ ] **MEDICINE AND MAN by Ritchie Calder.** Important medical events and discoveries from earliest times to the present, including the remarkable progress of our own generation. (#P2168)
- [ ] **RELATIVITY FOR THE LAYMAN by James A. Coleman.** An account of the history, theory and proofs of relativity, the basis of all atomic science. (#P2049)
- [ ] **THIS IS OUTER SPACE by Lloyd Motz.** A concise explanation of modern scientists' most recent discoveries about the nature of the universe. (#P2084)
- [ ] **THE ABC OF RELATIVITY by Bertrand Russell.** A clear, penetrating explanation of Einstein's theories and their effect on the world. (#P2177)
- [ ] **THE NATURE OF LIVING THINGS by C. Brooke Worth and Robert K. Enders.** A fascinating exploration of the plant and animal kingdoms, from algae to orchids, from protozoa to man. (#P2420)
- [ ] **THE WORLD OF COPERNICUS (Sun, Stand Thou Still) by Angus Armitage.** The biography of the great astronomer who established the general plan of the solar system accepted today. (#P2370)

THE NEW AMERICAN LIBRARY, INC., P.O. Box 2310, Grand Central Station, New York, New York 10017

Please send me the SIGNET BOOKS I have checked above. I am enclosing $_____(check or money order—no currency or C.O.D.'s). Please include the list price plus 10¢ a copy to cover mailing costs. (New York City residents add 5% Sales Tax. Other New York State residents add 2% plus any local sales or use taxes.)

Name_____

Address_____

City_____State_____Zip Code_____
Allow at least 3 weeks for delivery

MENTOR Books of Interest

☐ **MAN IN THE MODERN WORLD by Julian Huxley.** Stimulating essays on vital issues from Huxley's "Man Stands Alone" and "On Living in a Revolution."
(#MQ856—95¢)

☐ **HEREDITY, RACE AND SOCIETY (revised) by L. C. Dunn and Th. Dobzhansky.** Group differences, how they arise, the influences of heredity and environment.
(#MT883—75¢)

☐ **MAINSPRINGS OF CIVILIZATION by Ellsworth Huntington.** A penetrating analysis of how climate, weather, geography, and heredity determine a nation's character and history. Diagrams, maps, tables, bibliography.
(#MQ248—95¢)

☐ **THE ORIGIN OF SPECIES by Charles Darwin.** The classic work on man's evolution that revolutionized scientific and religious thinking from the 19th century onwards.
(#MQ503—95¢)

☐ **THE NEXT DEVELOPMENT IN MAN (revised) by Lancelot Law Whyte.** Art, politics, economics, and science are integrated into a "unitary" way of thinking in this stimulating book.
(#MP399—60¢)

☐ **AFTER THE SEVENTH DAY by Ritchie Calder.** A lively history of civilization in terms of man's genius for mastering his environment, and his terrible power for destruction. Photographs and drawings.
(#MT453—75¢)

☐ **FOLKWAYS by William Graham Sumner.** A classic study of how customs originate in basic human drives, this is an incredibly comprehensive research into the cultures of primitive and civilized peoples.
(#MQ766—95¢)

SIGNET Biological Science Books

☐ **MODERN GENETICS by Haig P. Papazian.** A layman's guide to the most basic of all the biological sciences, well illustrated with appropriate plates and figures.
(#Y3631—$1.25)

☐ **MAN IN THE WEB OF LIFE by John Storer.** An award-winning conservationist advises man of the necessity of a sound inter-relationship with all the aspects of his environment if he is to have a chance of survival in the future. (#P3664—60¢)

☐ **THE DAWN OF LIFE by J. H. Rush.** A lucid, absorbing explanation of the most recent and authoritative scientific thinking about the origin of life. (#T2192—75¢)

☐ **HEREDITY AND THE NATURE OF MAN by Theodosius Dobzhansky.** With deep concern for the humanity of mankind, a distinguished geneticist views the profound —and often controversial—social consequences of the study of genetics. (#P2837—60¢)
